Shirjeel
Tahir 1990

Longman Science 11–13 Series
General Editor: John L. Lewis O.B.E.

Also in this series:
Physics 11–13
Chemistry 11–13

BIOLOGY 11–13

P. J. Mawby M.A.
Head of Science, Cheltenham College
Formerly Head of Biology, Edinburgh Academy

M. B. V. Roberts M.A., Ph.D
Formerly Head of Biology, Marlborough College and
Cheltenham College

LONGMAN GROUP LIMITED
Longman House
Burnt Mill, Harlow, Essex CM20 2JE, England
and Associated Companies throughout the World

First published 1983
Second impression 1983
ISBN 0 582 31037 7

Set in 11 on 12pt Univers light, Monophoto 685

Printed in Great Britain
by Butler & Tanner Ltd Frome and London

Acknowledgements

We are grateful to the following for permission to reproduce photographs: Ardea London, cover (Pat Morris); Aquila Photographics, pages 72 above right (T. Leach), 93 and 101 (M. C. Wilkes), 125 (R. R. Page), 153 above left (E. K. Thompson) and 276 left (T. Leach); Biofotos, pages 7 left and above and below right, 10 centre and below right, 24 left, 103, 121, 146 and 188 centre left; Biophoto Associates, pages 50, 111, 115, 129, 132 left (T. Norman Tait), 152, 155, 215, 230 and 264; John Cleare/Mountain Camera, page 180; Eric and David Hosking, pages 24 centre, 72 above and below left, 87 above and 273; Alan Hutchison Library, page 165; Frank Lane, pages 7 centre right (Basil Collier), 10 left (Leonard Lee Rue), 10 above right (G. E. Kirkpatrick), 72 below right (Leonard Lee Rue), 126, 131 (Lynwood M. Chace), 132 right (Ben M. Knutson), 153 above right (Leonard Lee Rue), 153 below (Karl H. Maslowski), 163 (A. J. Roberts), 174 (S. Jonasson), 188 above left (Alaska Pictorial Service), 188 above right and below left (Leonard Lee Rue), 188 centre right (Walther Rohdich), 188 below right (Robert C. Hermes) and 276 right (Irene Vandermolen); London Scientific Fotos, page 87 below; Pace, page 257; Press-Tige Pictures, pages 41, 77, 170, 216, 258, 269, 270 and 271 above; Royal Dental Hospital of London School of Dental Surgery, page 265; Science Photo Library, page 263 (Tony Brain); Seaphot/Planet Earth Pictures, pages 178 (Rod Salmon), 179 (John Lythgoe), 183 (Peter David), 188 above centre (Vincent Serventy) and 188 below centre (Herwarth Voightmann); John Topham, pages 149 (Associated Press) and 271 below (IPA); Wellcome Institute, page 59.

For permission to redraw copyright material:
Appendix 3, pages 282–8, Heinemann Educational Books,

Clarke et al, *Biology by Inquiry Book 1*; fig 19.11, page 274, Macdonald Educational, Gilman, *Urban Ecology*; fig 18.4, page 255, McGraw-Hill Book Company, Demarest, *Conception, Birth and Contraception*; fig 18.8, page 260, Oxford University Press, Harrison et al, *Human Biology*; fig 10.7, page 144, Pelican Books, Fogg, *The Growth of Plants*.

We are grateful to the following for permission to reproduce copyright text material:
Common Entrance Publications Ltd and Common Entrance Examinations Science Advisory Panel for Q B1c, B2b & c November 1975; Q A1a,b,c & d, B1a & b June 1976; Q B1 February 1977; Q A2, B1 and B2 June 1977; Q B1, B2a,b,c & d November 1977; Q A3, B1, B1a & B2 February 1978; Q A2 June 1978; Q B1 b–g November 1978; Q A2 & B1 February 1979; Q B1 November 1979; Q A2 February 1980; Q B3 1980 Specimen Paper; Q A2 & A3 June 1981; Q A2 & B1 November 1981; Q A1 & B3 February 1982 and Q B1 June 1982; I.L.E.A. for two questions adapted from Curriculum Development Team, *Insight to Science* Project; London Editions Ltd and Thomas Nelson and Sons Ltd for a simplified extract from *The Amazing World of Animals* edited by Sir Peter Scott; the author's agent for a simplified extract from *Across the Russias* by John Massey Stewart; Oxfam and UNICEF for an adapted extract from *Together for Children* 1979.

Contents

Preface

This book is intended to serve as an introduction to biology for 11–13 year olds. Our aim has been to lay the foundations of the subject in a clear and, we hope, interesting way.

Children discover things about themselves and their surroundings by question and experiment. For this reason we have included numerous experiments and questions in this book. The experiments are of a type that can easily be carried out with the minimum of equipment. The questions are of two kinds: some are intended to be a basis for class discussion; others are more suitable for homework.

Plainly children cannot discover everything for themselves; much of their knowledge comes from reading or listening to the teacher. This method of acquiring knowledge should not be despised, particularly if it prompts children to ask further questions and opens the way for discovery. We certainly hope that the information provided in this book fulfils this aim. Because we attach considerable importance to reading, we have included selected passages of background reading at the ends of some of the chapters.

It is not easy to produce a book that is suitable for a wide range of ability in the 11–13 age group. If we have succeeded it is largely because of the help we have received from many friends and colleagues. We would particularly like to thank Mr Richard Balding, Mrs P. M. Cullen, Mr Alec Porch and Mr Robert Powell, all of whom have read the manuscript and made many valuable suggestions. We are also grateful to Mrs Megan Woods for typing the manuscript.

Finally we owe grateful thanks to John Lewis, General Editor of the Longman *Science 11–13* Series, for his helpful advice and his never failing enthusiasm and encouragement.

Cheltenham 1982

P. J. Mawby
M. B. V. Roberts

Chapter 1 # Where animals and plants live

What is biology?

Biology is the study of living things and a person that studies living things is called a *biologist*. One of the first questions a biologist asks about an animal or plant is, 'where does it live?' The place in which an animal or plant lives is called its *habitat*.

This chapter is about habitats, but we shall also look at the surroundings of living things. An animal or plant's surroundings make up its *environment* and every habitat must provide the animals and plants which live there with the right environment.

What is your environment now?

At this moment you are probably sitting in a classroom or laboratory. For the time being this is your habitat, and it provides you with a suitable environment in which to live and work: light to read by, air to breathe, a hard surface to support your weight, and perhaps a heater to keep you warm. We call these the *physical features* of your environment.

However, you are also affected by your classmates and your teacher, and also by other forms of life round about you, such as bacteria in the air. These make up the *biological features* of your environment.

Experiment 1.1 Measuring the physical features of your environment

Two of the most important physical features of your environment are light and temperature. Can you suggest a way of measuring these two features in the room in which you are working?

Measure the temperature in different parts of the room and work out the average. (To do this, add up the readings and divide by the total number of readings.) If the necessary equipment is available, measure the light in different parts of the room as well.

Questions for class discussion

1. Make a list of other physical features of your environment besides the ones you have measured. Which ones are essential for life and which might be changed by your presence in the room?

2. A pupil measured the temperature of the environment in six different places and obtained the following readings: 13°C, 17°C, 18°C, 21°C, 15°C, 12°C.
a. By how many degrees did the temperature vary?
b. What was the average temperature?
c. Suggest reasons for such variable readings.

3. Four different pupils measured the temperature of their classroom at three different times of the day and got these results:

9 a.m.   13.2°C
12 noon 17.5°C
6 p.m.   16.9°C

Andrew recorded the average as 16, Gill as 15.9, Peter as 15.866 and Diane as 15.

Whose answer is best, and why? What is wrong with the others?

Artificial habitats

One of the best ways of learning about habitats is to make an artificial one. For example, you could set up a large bottle containing various animals that live on the ground: we call this a *terrarium*. Or you could make a *wormery* out of two plates of glass with soil in between, or an ants' nest out of plaster of Paris. You will find instructions for setting up such habitats in the books listed in the Appendix.

One of the simplest habitats to make and maintain is a fresh-water *aquarium*, and it has the advantage that you can observe it at frequent intervals and make measurements easily.

Experiment 1.2   Setting up an aquarium

You will need a container: a bowl, deep tray or tank will do, but it is best if it is wide and fairly shallow. Wash it thoroughly, then add a few centimetres of sand or gravel. Carefully pour in some stream or pond mud and spread a few large stones round the bottom. Finally add water, preferably river or pond water, pouring it over one of the stones to reduce disturbance of the mud. *The depth of the water should not be greater than half the width of the open surface of the container.*

If the aquarium is to be left outside, cover it with perforated zinc or fine-mesh chicken wire to keep out dead leaves. If it is to be kept indoors, cover it with a sheet of glass or wire-reinforced plastic and make sure it is well lit: a fluorescent strip is best if there is insufficient daylight.

When the mud has settled, make a rough sketch-map of your aquarium as seen from above. You can make your map more accurate if you attach strings 10 cm apart across the top of the aquarium dividing it into squares of equal size (a *grid*) as shown in Fig 1.1.

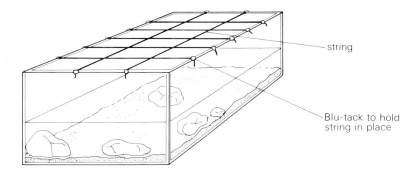

string

Blu-tack to hold string in place

Fig 1.1 An aquarium grid

Questions for class discussion

1. Why do you think the water in your aquarium must not be too deep?

2. What are the advantages of covering an indoor aquarium with glass or plastic?

3. Why is it necessary to light the aquarium well, and why is a fluorescent strip a good source of light?

4. As your aquarium settles down, various changes take place. What do you think will happen to it
a. in the first 24 hours
b. if it is left untouched for several months?

Life in an aquarium

We refer to living things, both animals and plants, as *organisms*. Many organisms found in fresh water are suitable for keeping in an aquarium, and your next job will be to find some. But first, think about the number and types of organisms which you want. Apart from availability and size, what should you consider? Where will you get them from? What sort of equipment will you need? Your teacher will help you to answer these questions.

Experiment 1.3 Stocking up your aquarium
Fig 1.2 shows some of the organisms which you might put into your aquarium as a start. They are common in ponds and streams.

Take a net, a bucket and any other equipment you think you may need, and go out to a nearby pond, lake, river, stream, ditch – even a gutter or water butt might do – and find as many of the organisms as you can. Bring them back to the laboratory.

Before you put any organisms into your aquarium, bear in mind the following important points:

1. Try not to disturb the mud when you put in the organisms. Plants with roots should be weighted down with stones and carefully lowered into the mud.

2. Many small organisms are better than a few large ones, and the more different kinds the better. Why?

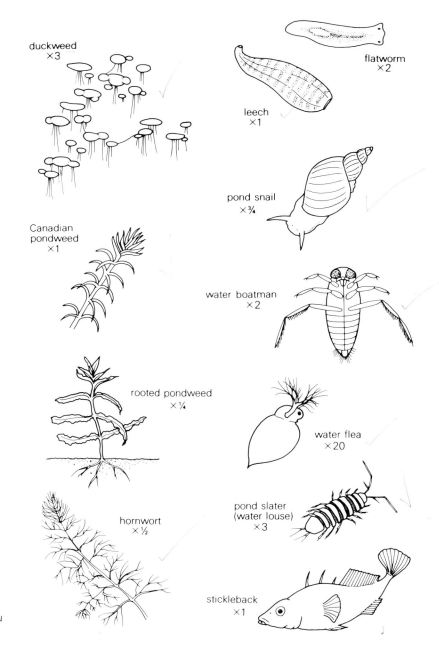

duckweed
×3

flatworm
×2

leech
×1

Canadian
pondweed
×1

pond snail
×¾

water boatman
×2

rooted pondweed
×¼

water flea
×20

hornwort
×½

pond slater
(water louse)
×3

stickleback
×1

Fig 1.2 Some organisms that you
might put in an aquarium

5

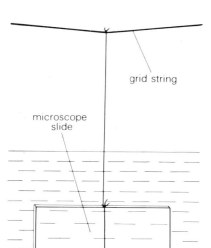

grid string

microscope
slide

Fig 1.3 A method of collecting
microscopic organisms in an
aquarium

3. Make sure you put in plenty of plants and *not* too many animals. Why do you think this is important?
More animals can be added if extra air is provided by an aerator.

Make a list of the animals and plants which you put into your aquarium and note how many there are of each type. Indicate on your map (Experiment 1.2) the positions of the rooted plants.

Finally hang some microscope slides from the grid strings as shown in Fig 1.3: with luck, some small organisms will become attached to the slides and grow on them.

Questions for class discussion

1. If your aquarium is kept indoors, the water level will gradually fall. Why does this happen? Which of the following should you add to keep the conditions inside constant: tap water, rain water or pond water?

2. A cold water aquarium does not need much attention, but it is advisable to check it from time to time and remove any dead animals or plants. Why should this be necessary in an aquarium when it does not happen in a pond or river?

Natural habitats

There are all sorts of natural habitats, some of which are shown in Fig 1.4. Examine them carefully and write down one feature which distinguishes each habitat from all the others.

The name given to all the different kinds of living organisms in a habitat is a *community*. One of the most familiar examples of a community is a lawn. Although it consists mainly of grass, even some of the best-kept lawns contain 'weeds' such as the ones illustrated in Fig 1.5.

Fig 1.4 Four natural habitats

(b) a sand dune

(c) a meadow

(a) a wood

(d) moorland

plantain

daisy

buttercup

clover

dandelion

Fig 1.5 Some weeds found in a lawn

Experiment 1.4   Mapping a plant community

Visit a lawn or playing field near your school and make a map of the plant community as follows:

1. First set out a grid of strings one metre apart, held in place by pegs. Each pupil should take one square metre to map (Fig 1.6(a)).

2. Outline the areas occupied by each type of plant. Do not show individual plants unless they are large and isolated from others of the same kind (Fig 1.6(b)).

3. Use a key to show which plant is which (Fig 1.6(c)).

When you get back to the classroom, you can put all the maps together to make a *habitat map* of the whole lawn.

Questions for class discussion

1. Your map shows only the plant communities in the lawn. What animals might live in the habitat or might visit it occasionally? How might these animals affect the plants?

2. Are certain plants more common in some areas than in others? If so, suggest possible reasons. Here are two questions that might help:

a. Are there any hollows in the lawn? Does a particular plant occur more frequently in the hollows than else-where? If so, which plant is it? What advantages might there be to living in a hollow?

b. A lawn is a comparatively simple habitat to map. Why is this?

Seasonal changes in natural habitats

Write down the changes that occur in a lawn between one season and another. In fact the changes are not very great. Another habitat which shows relatively little seasonal change is the seashore: you can see this for yourself if you

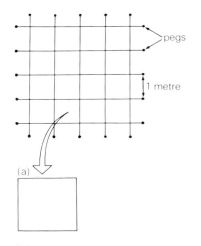

(a)

(b)

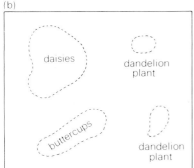

daisies

dandelion plant

buttercups

dandelion plant

(c)

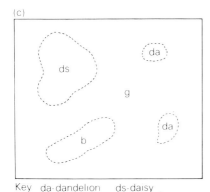

ds

da

g

b

da

Key   da-dandelion   ds-daisy
      g-grass   b-buttercup

Fig 1.6 Mapping a plant community

visit a rocky shore and examine a rock pool at different times of the year. Even if you live in the centre of a city there are plenty of habitats with interesting animals and plants which you can examine at different times of the year: roofs, gutters, walls, and loose stones are some examples.

If you visited a wood in the summer, and then returned to the same wood in winter, you would find that many changes had taken place. Here are three reasons why the animals and plants found in a wood in Britain change in autumn: *or, winter sleep*

1. Hibernation. Some animals that find it difficult to feed and keep warm through the winter become less active in the autumn and may go to sleep. An example is the squirrel., *frog,*

2. Migration. Many birds, and some other animals, travel long distances in the autumn to warmer places. They do this in order to avoid the harsh conditions of winter. Can you name some of them? Where do they go?

3. Leaf fall. In temperate countries many trees shed their leaves in the autumn. This allows more light to reach the ground. It also cuts off the food supply for animals that feed on living leaves, but increases the food supply for organisms that feed on dead leaves.

Questions for class discussion
1. Make a list of the main changes that occur in a wood in the spring.

2. Habitats and communites abroad, for example a desert or a tropical rain forest, wil differ considerably from those found in Britain. What special features might these habitats provide, and why are they particularly suitable for the organisms that live there? To what extent do they change with the seasons?

3. Return to your aquarium. What changes can you see? Try to explain why the distribution of some organisms has changed while that of others has stayed the same.

Fig 1.7 In what sort of environment would you expect to find these animals and plants?

(b) polar bear

(d) hawthorn

(a) cactus

(c) sea lion

Homework assignments

1. Some pupils carry out a project to find out what happens to fallen leaves in an oak wood. They cordon off an area of four square metres with chicken wire, and they measure the depth of the dead leaves at regular intervals. Their results are shown in Table 1.1.
   a. Suggest reasons why the depth of dead leaves gets less.
   b. Why do you think the depth remains unchanged between the middle of January and the middle of February?

2. Carefully examine the animals and plants shown in Figure 1.7. Describe the sort of environment in which you would expect to find each one, giving reasons for your choice.

3. You are asked to find out as much as possible about a local pond and its plant and animal life. How would you set about the task of making a map of the pond and its immediate surroundings?

4. Suppose you visited a woodland habitat in late summer. Describe *three* differences you would expect to observe when revisiting the same habitat in the middle of winter.

5. You are supplied with a glass-sided aquarium about 30 cm × 15 cm × 15 cm high, some sand and small stones, Canadian pondweed and some duckweed. Describe the stages by which you would make the aquarium ready to receive pond animals.

Table 1.1

Date	Depth
1 December	16 cm
15 December	13 cm
1 January	10 cm
15 January	9 cm
1 February	9 cm
15 February	9 cm
1 March	6 cm
15 March	5 cm
1 April	5 cm
15 April	3 cm

Background reading

The animal life of Siberia
Polar bears and camels, walruses, tigers and Japanese cranes . . . and birds and mammals only recently thought to be extinct. Greater Siberia has an extraordinary range of fauna in its five million square miles, fifty-three times the size of the United Kingdom. Across it stretch three belts of vegetation: tundra, taiga (or coniferous forest) and steppe.

The tundra zone is now shrinking fast as a result of the increase of far northern temperatures over the past fifty

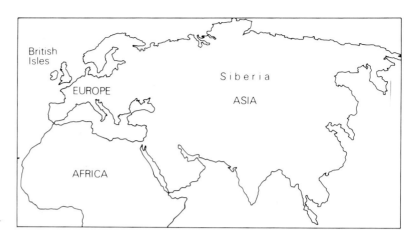

Fig 1.8 Siberia

years. Forests have advanced north into the tundra at the rate of almost half a mile every year. In the summer of the midnight sun the tundra comes to life. The Arctic fox, snowy owl and its favourite prey, the lemming, are joined by animals migrating north – brown bear for instance, blue fox, reindeer and wolf, all escaping from the so-called 'scourge of the taiga', the blood-sucking mosquitoes, midges, gnats and horse-flies which accompany the summer's heat and can drive both man and animals demented.

The taiga, by far the largest vegetational belt of Greater Siberia, supports a far richer fauna than the tundra: about 150 species of mammals and more than 200 species of birds. The taiga's mammals incorporate the elk, reindeer, roebuck and the beautiful (protected) maral deer, the wolf, fox, glutton (or wolverine), lynx and the master of the taiga, the seven to eight foot brown bear.

South of the taiga most of the original steppe vegetation has been ploughed up. But in its last vestiges, hosts of butterflies flutter among the flowers and scorched grasses, and birds abound. Abundant too are the rodents and burrowing animals but Przhevalsky's horse is seen no more and man has killed off the aurochs.

Certainly the most remarkable aquatic habitat is Lake Baikal, the world's oldest and deepest lake and the greatest body of fresh water on Earth. Its water is by nature exceptionally pure thanks to a tiny underwater creature, the dotted cymatoa, which lives on the micro-organisms normally responsible for 'water bloom'. It filters the lake water through its tendrils, cleaning the entire surface layer several times a year, so that the water flowing out of the lake is actually much cleaner than that flowing in. Largest of all the lake's fauna is the freshwater seal, which numbers about 35,000 and lives in winter beneath the ice, breathing through ice-holes and bearing its young in lairs on the ice surface.

The most famous of all Siberia's fauna, however, has been extinct for just over 10,000 years: the mammoth. But that, as they say, is another story.

(From *Across the Russias* by John Massey Stewart, published by Harvill.)

Questions
1. Using a dictionary if necessary, give the meaning of each word underlined in the passage.

2. To what seasonal changes does the author refer?

3. Why do you think the taiga has a richer fauna than the tundra?

4. Why is the water that flows out of Lake Baikal cleaner than the water flowing in?

5. What kind of animal was a mammoth?

Summary
1. A *habitat* is a place in which organisms live, and it provides the organisms with a set of conditions which together make up the environment.

2. The *environment* is composed of physical and biological features.

3. The various organisms which live in a habitat make up a *community*.

4. Habitats and communities are always changing, for example, with the seasons. Some change more than others.

5. Seasonal changes bring about leaf fall, hibernation and migration.

Chapter 2

Observing and measuring living things

Similarities and differences

How *different* are you from the other members of your form? Which would you say are the main features by which your teachers and school friends recognise you?

How *similar* are you to the rest of the form? How many likenesses can you think of? If you list the similarities and differences, which list is longer?

Now consider the similarities more carefully. How many of the features are exactly the same (*identical*)? Perhaps you have included age in your list. Are any of your form *exactly* the same age as you? Your list may have included some characteristics of the human body: two eyes, one nose – but the eyes will be of different colours and the noses will differ in length and shape.

Only identical twins have identical features. This chapter deals with the differences between individuals, starting with *you*.

Experiment 2.1   Measuring differences in human height

Height is not difficult to measure, but you must follow the instructions carefully to obtain an accurate measurement.

1. Arrange yourselves into groups of three.

2. Fix a pair of metre rules end to end to a wall or door, making sure that the centimetre scale on each rule runs from 0 at the bottom to 100 at the top.

3. A member of each group is chosen to be the first person to be measured (called the *subject*). The subjects remove their shoes and stand with their backs to the rules with their feet together. (Fig 2.1.)

Fig 2.1 Measuring height

4. The measurer places a book at right angles to the metre rule across the top of the subject's head.

5. A third member of the group checks that the subject is standing correctly (heels on ground, shoulders back) and ensures that the book is level.

6. The subject walks away and the measurer keeps the book in position. The height opposite the lower edge of the book is read to the nearest millimetre (the smallest division on a metre rule).

7. Write down the subject's height in centimetres. Do not forget to add 100 to the reading to allow for the lower rule.

8. Now take it in turn to measure the other members of your group.

9. Make a list of the names of the pupils in your form and write down their heights in a column beside their names. You have now expressed your results in a *table*. Do not forget to write down the *units* you have used: in this case *centimetres*. Part of your table might look like Table 2.1.

Table 2.1

Name	Height (in centimetres)
John Smith	121.3
Susan Jones	153.1
Peter Brown	130.5

10. Underline the height of
a. the tallest person
b. the shortest person in the form. Subtract b. from a. to obtain the height *range*.

11. Add together the heights of all the pupils and divide by the number measured to obtain the *average* height. Write down the range and the average beside your table; together they give useful information about the differences in height of the people in your class.

12. Finally, it is important to record the *accuracy* of your measurements. If you followed the instructions carefully, and if you were able to read the rule to the nearest millimetre, then your measurements should have been within half a millimetre of the 'correct' height. Write this beside the table in the following way:

Accuracy = ± 0.5 mm

Drawing a bar graph

Measurements such as the ones you have taken are usually shown in the form of a *bar graph* (also called a bar chart). It is more convenient than a table of results, especially if a large number of measurements have been obtained. It also makes it easier to compare the measurements at a glance.

1. You have already found the height range for your form in Experiment 2.1, part 10. Divide the range into 5 to 8 equal parts. We call these *classes*. For example, if the smallest pupil is 121.3 cm and the tallest is 153.1 cm, the height classes could be 121–125 cm, 126–130 cm, 131–135 cm, 136–140 cm, 141–145 cm, 146–150 cm and 151–155 cm.

2. Make a table with the height classes listed on the left-hand side.

3. Put a tick beside the height class for each pupil's measurement. The pupil whose height was 153.1 cm would go in the class 151–155 cm. A height which is exactly between two height classes such as 130.5 cm should be placed in the upper class 131–138 cm.

4. Add up the number of pupils whose heights fall within each height class, and write the total for each height class in the right-hand column of your table. Your table should look something like Table 2.2.

5. Take a piece of graph or squared paper and draw a horizontal line about 3 cm from the lower edge. This is the *horizontal axis* and will represent the height classes. Now draw a vertical line about 3 cm from the left-hand

Table 2.2

Height class (cm)	Pupils in each height class	Total in height class
121–125	√	1
126–130	√ √	2
131–135	√ √ √	3
136–140	√ √ √ √ √ √	6
141–145	√ √ √	3
146–150	√ √	2
151–155	√	1
Total number of pupils 18		

margin of your paper. This is the *vertical axis* and will represent the number of pupils in each height class.

6. How many height classes did you have? Our example had seven height classes. Divide the horizontal axis into sections of equal size, for example 2 cm each. Label each section with its range.

7. What was the largest number of pupils in a single height class? In our example there were six people in the 136–140 cm class. Divide the vertical axis into equal-sized sections to take the largest number of measurements in the right-hand column of your table. The axes of your graph should now look something like Fig 2.2.

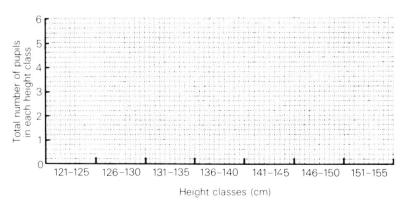

Fig 2.2 Drawing the axes for a graph

8. Now draw a series of columns, one in each height class. The columns should represent the figures in the final column of your table. If the total in height class 121–125 cm is 1, then the column will go up to the 1 on the vertical axis.

What does a bar graph tell us?

How many pupils did you measure and include in your bar graph? The number of people measured is called the *sample*. If the sample is large enough we expect the highest point to be approximately half way along the height range – quite close to the average height. This is called a *normal distribution*. You can see an example in Fig 2.3.

The differences that you observe between members of your form are referred to as *variation*. If your graph of variation in height does not look like a normal distribution you should try to suggest reasons. The questions that follow may help you.

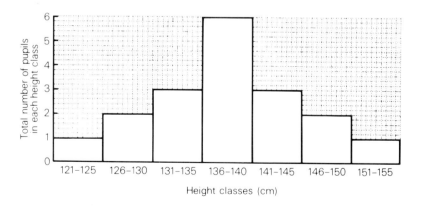

Fig 2.3 A bar graph based on the measurements in Table 2.2

Questions for class discussion
1. Find and mark the point on your graph corresponding to
a. the mid-point of the height range
b. the average height
c. the highest point (*peak*) of your graph.

2. If the heights of the pupils in your class give a normal distribution, then a, b and c in question 1 will be roughly the same. How close are they?

3. Suppose you were able to measure twice as many pupils of your age. What difference would you expect it to make to your graph?

4. What would be the effect on your graph of including *all* the pupils in your school?

5. If your measurements hac been made on a group of boys or a group of girls only. would there have been any differences?

6. Suppose the measurements had been taken less accurately (say $\pm$ 1 cm). How would the graph be affected?

7. Which of the following can be found from your bar graph *alone*?

a. The range of heights
b. The average height
c. The size of your sample
d. The height of an individual person in your form
e. The age of an individual if you know his height

Experiment 2.2　Finding some other differences

Here are some other differences which you might study. You may be able to think of some more. Record your own details in your notebook, then collect together the results from the whole form.

Mass

A pair of bathroom scales is suitable, although you may have to convert the mass from stones and pounds into kilograms. Proceed as follows:

1. Adjust the scales to zero.

2. Remove your shoes and jacket.

3. Stand in the centre of the step.

4. Check the units very carefully. Metric scales will give the mass in kilograms (kg), but older ones will be in stones and pounds.

Arm-span

Any smooth vertical surface will do to measure your arm-span; alternatively a laboratory bench could be used.

1. Remove your jacket or cardigan.

2. Stand against the blackboard or wall, or lean over the bench.

3. Stretch out both arms as widely as possible (Fig 2.4).

Fig 2.4 Measuring arm span

4. Mark the positions of the fingertips with chalk.

5. Measure the span with a rule or tape.

Finger prints

Your finger print pattern can be seen with a hand lens, but it is made clearer if the finger is inked on a stamp pad, and pressed against a sheet of paper (Fig 2.5).

1. Place a drop of ink on the pad and allow it to soak in. (Some pads do not need to be inked.)

Fig 2.5 Finger prints

2. Roll your finger lightly on the pad. Then roll it on a piece of white paper.

3. Repeat on another part of the paper, without re-inking your finger, until the ink runs out.

4. Select the clearest print. Examine it carefully under a hand lens and compare it with the prints of the rest of your form.

Questions for class discussion

1. Finger prints are often used in evidence when a person is suspected of having committed a crime.
a. List as many differences as you can between the finger prints of the members of your form. Are any of the prints the same?
b. Finger prints would be difficult to measure, but they can be divided into classes. Suggest how this might be done and estimate how many members of your form would be included in each class.

2. Look carefully at the height, mass and arm-span measurements of the members of your class. Has the tallest pupil got the largest arm-span? Is the smallest also the lightest? Can you see any other connections?

Homework assignments

1. A school-friend knows nothing about graphs. Try to explain very simply what is meant by a normal distribution, using the words *sample*, *average* and *range* in your explanation.

2. The height of each 11-year-old pupil in a London school was measured. The results are shown in Table 2.3.

Table 2.3

Height class (cm)	Number of pupils in each height class
121–125	1
126–130	10
131–135	18
136–140	17
141–145	9
146–150	1

a. Show these measurements on a bar graph.
b. Show on your bar graph
   (i) the *average* height (134.0 cm)
   (ii) the middle of the height *range*
   (iii) the height class containing most pupils.
3. Compare the results given in question 2 with the results which you obtained in Experiment 2.1. List the differences and try to suggest some reasons for them.

Experiment 2.3 Using a hand lens

Suppose you want to distinguish between the individuals in a group of small animals. If they are too small to see clearly or if you have to examine them in detail, a *hand lens* is very useful.

These lenses usually make an object look five to ten times as large as their life size. We say that their *magnification* is 'times 5' or 'times 10' and we write this as × 5 or × 10. This is called the *scale*. Practise using a hand lens by looking at a coin, stamp or piece of microfilm.

1. Hold the object in one hand under a good light.

2. Hold the lens in your other hand *close to your eye*.

3. Move the object slowly up towards the lens, starting about 30 cm away, until it just becomes clear and sharp. (Fig 2.6.)

4. Look at the details on the stamp or coin (for example, the decorations on the crown).

5. Work out the approximate magnification using the perforations on the edge of the stamp, or the number of letters visible on the microfilm.

Fig 2.6 Using a hand lens correctly

Experiment 2.4 Recording what you see

Recording your observations by means of sketches or drawings is a very useful skill. You do not have to be a good artist, but you do have to train yourself to observe carefully.

You need a small animal such as a woodlouse, a hand lens, a ruler, a soft rubber, a sharp HB pencil and a sheet of

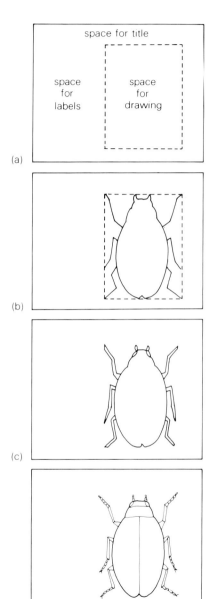

(a)

(b)

(c)

(d)

Fig 2.7 Recording observations by drawing

plain paper. Proceed as follows:

1. Measure your paper and decide how large your drawing is to be when completed. Make the drawing as large as you can, but allow space for a heading and some labels.

2. Measure the length and the breadth of the specimen and decide how many times larger your drawing will be.

3. Now use the ruler to indicate, with very faint pencil lines on your paper, the vertical and horizontal edges of your drawing. (Fig 2.7(a).)

4. Now sketch the *outline* of the specimen, very softly. Make corrections if necessary until you are satisfied with the outline. (Fig 2.7(b).)

5. Sharpen your pencil again if necessary and go over the final sketched outline to give a firm, clear, single outline. Avoid the common mistake of giving the specimen a fuzzy outline; try to go right round corners with a single movement of the pencil. Rub out all the sketch lines once the outline is complete. (Fig 2.7(c).)

6. Fill in the details, if necessary using the hand lens to get them accurate in shape and position. (Fig 2.7(d).)

7. Write the names of the main parts of the specimen (for example, head, feelers, legs) alongside the drawing and use a ruler to join the name to the drawing. This is called *labelling* your drawing.

8. Do not forget to add:
a. The name of the specimen.
b. The place where it was found.
c. The magnification of your drawing. This can be written as × 10 (refer back to instruction 2) or by giving an indication of the natural size such as 'length = 7 mm'.

Questions for class discussion or homework
1. The photographs in Fig 2.8 all show common objects reduced in size. Use a hand lens to identify the objects. Make sure the lens is close to your eye and the book is in a good light.

Fig 2.8 What are these common objects?

 (a)

 (b)

 (c)

2. The magnification or scale is often used to describe the actual size of an object from a drawing, map or plan. Can you answer the following questions about scale?

a. An architect's drawing shows my house 20 cm wide. What is the actual width if the drawing's scale is 1:100?

b. The famous cannon in Edinburgh Castle called Mons Meg is 4 metres long. How many centimetres long would a 1:40 kit model be? (1 metre = 100 cm.)

c. The Ordnance Survey produces maps of the whole of Britain based on a 1:50 000 scale. How long would the largest bridge span in the world, the Humber Estuary Bridge, appear on the map? Its actual length is 1.4 km. Give your answer in centimetres (1 km = 1000 metres).

3. The lens I normally use is marked × 10. When I use it to look at a ladybird 3 mm long, how large does the ladybird appear to be? My drawing of the ladybird is five times as large as it appears under my lens. Will it fit a piece of paper the size of this page?

4. A student studying variation in size within a single species of snail, collected a large number of snails from a railway embankment and arranged them into classes according to their length (Table 2.4).

Table 2.4

Length classes (mm)	Number of snails in each length class
0– 2.5	0
2.6– 5.0	35
5.1– 7.5	175
7.6–10.0	185
10.1–12.5	190
12.6–15.0	55
15.1–17.5	5
17.6–20.0	0

a. Make a bar graph showing these results.
b. What type of distribution does the graph show?
c. What is the most common length class?
d. Why is it not possible to work out the average snail length from the information given?
e. Suggest a reason for the absence of snails in the range 0–2.5 mm and a reason for the absence of snails longer than 17.5 mm.

5. The graph in Figure 2.9 compares the average heights of boys in the United Kingdom aged between 10 and 19 in 1958 and 1878. Examine it carefully before attempting the questions.

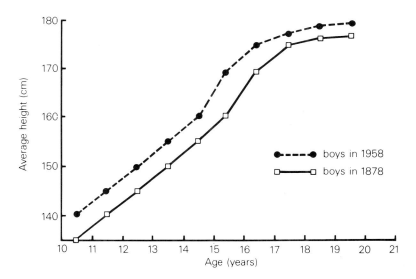

Fig 2.9 Average heights of boys in the U.K.

a. What is the average difference in height between 1958 boys and 1878 boys?
b. At what age is there the greatest difference in height between 1878 boys and 1958 boys?
c. In which year of the boys' life was growth most rapid: (i) in 1958   (ii) in 1878?
d. What reasons can you think of to explain the greater height and earlier growth of 1958 boys? Do you think the changes are likely to continue?

6. Refer to the picture in Fig 3.5 on page 30.
a. Which drawing is less than life size?
b. Which is the largest animal shown?
c. What is the length of the water beetle?

Summary

1. Organisms of a single type show differences from each other; careful measurement and observation is needed to detect this *variation*.

2. Measurements can be analysed by calculation of the *average* and *range*.

3. The measurements can also be shown as a *bar graph*. If the sample is large enough, a *normal distribution* is frequently obtained.

4. Small differences can be observed and measured using a *hand lens*. The observations are recorded by making a careful drawing. The *magnification* or *scale* of the drawing is always noted.

Chapter 3 **Sorting and naming living things**

Sorting living things

If you have a collection of any kind, you will know how difficult it can be to sort out the items. In this chapter you will learn how living things are arranged in groups or sets. Biologists call this process *classification*.

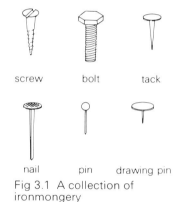

screw   bolt   tack

nail   pin   drawing pin

Fig 3.1 A collection of ironmongery

Experiment 3.1   Putting things into groups
Your teacher will give you six items to classify. Suitable things might include coins, laboratory glassware (beakers, flasks, test tubes etc.) or ironmongery (nails, screws, bolts etc.)

1. Give each item a letter or name so that you can refer to it easily. Write the letter or name, with a sketch, in your notebook (Fig 3.1).

2. Look for one feature that some of the items possess but the rest do not. Separate the items into two piles on your desk and record the feature you used in your book (Fig 3.2).

Fig 3.2 Separating the collection into two groups

items with a screw thread   items without a screw thread

3. Now choose a feature that separates the items in *one* of the piles. Move them apart and continue until each item is on its own.

4. Do the same with the other piles and record what you have done. The record in your book should look like Fig 3.3.

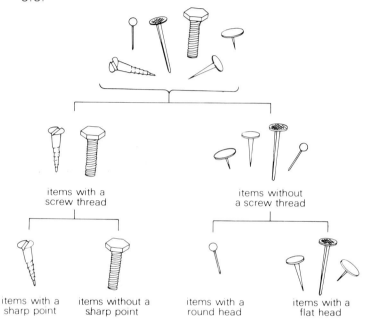

Fig 3.3 Separating the collection into four groups

Questions for class discussion

1. Look at Figure 3.3 and find which items have not been separated. What feature could you use to separate these items?

2. The ironmongery items in Figure 3.3 differ in size but they have not been separated according to their size. Why not?

3. Most stamp collections are arranged by country. What other methods could be used to classify collections of stamps? Can you think of one advantage and one disadvantage of grouping a stamp collection by colour?

Homework assignments

1. The six shapes in Fig 3.4 have been partly separated into groups for you. Copy down the scheme, writing in the features that were used to separate them, and then complete the classification.

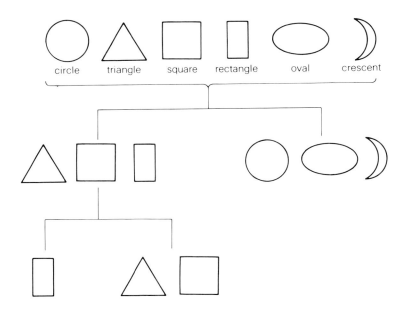

Fig 3.4 Which features have been used to separate these shapes into groups?

2. Use the method described in Experiment 3.1 to separate the following items into groups:
a. knife, fork, teaspoon, dessertspoon, tablespoon
b. cow, sheep, pig, horse, hen
c. bicycle, car, bus, train, aeroplane

Experiment 3.2　Classifying living things

There are almost a million different kinds of animals and a third of a million different kinds of plants known and named. You will be given, or asked to collect, specimens of 6–10 living things, for example, twigs, fruits or leaves from different kinds of trees, or shells of different animals.

1. Sketch each organism and give it a name.

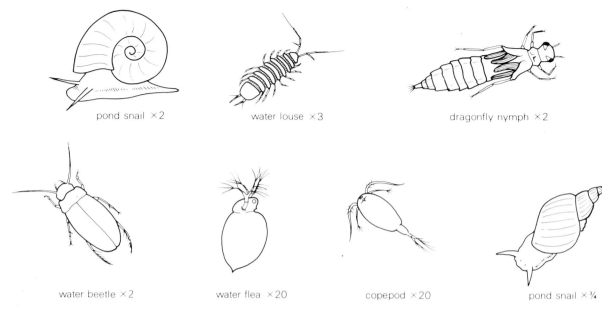

pond snail ×2   water louse ×3   dragonfly nymph ×2

water beetle ×2   water flea ×20   copepod ×20   pond snail ×¾

Fig 3.5 Some aquarium animals

2. Use the method described in Experiment 3.1 to separate them into two groups, then divide each group until each organism is on its own.

3. Record what you have done, as in Fig 3.3.

Questions for class discussion

1. Exchange your classification and collection of specimens with your neighbour's. Compare different methods of classification. Are some methods better than others? If so, can you decide why?

2. How useful would your classification be
a. in a year's time?
b. at a different time of year?
c. with other kinds of specimens?

Homework assignment

Fig 3.5 shows some animals which you might find in your aquarium.

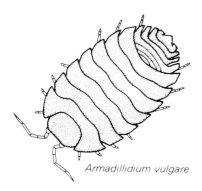

Armadillidium vulgare

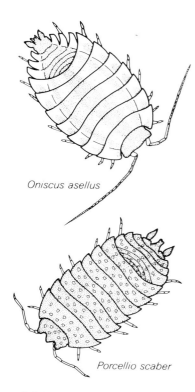

Oniscus asellus

Porcellio scaber

Fig 3.6 What is the common name for these animals?

a. Sort the animals into three pairs according to obvious external features. Which is the odd-man-out?
b. Make a classification of them, similar to the classification in Fig 3.3.

Naming living things

Each particular kind of animal has its own common name. You will be familiar with many of them, for example, robin, snail and human. What do you call the little animals which are drawn in Fig 3.6?

The most common names are probably 'woodlouse' and 'slater' but you may be suprised to learn that in Devon alone there are over 34 different names for these animals. To make things even more complicated, the word 'woodlouse' is used to refer to at least 38 different kinds of animal. Three kinds are shown in Fig 3.6; how many differences between them can you see?

Each different kind of animal and plant is called a *species* and scientists have given each species a different scientific name; no other species has the same name. The names of each of the three species of woodlice are written below the drawings in Fig 3.6. You will notice two things about these names.

First, they are in Latin. All the scientific names given to living organisms are in Latin because this was the international language of scholars in the 18th century when the system of naming we use today was begun.

Second, each organism has two names. The first name tells you the group of organisms, or *genus*, to which it belongs and the second name identifies the species. As biologists collect organisms more widely and study them more closely, new species are being discovered and named each year. Each new discovery has to be examined and compared with species which already have names. The biologist looks for similarities and differences so that the specimen can be placed in the right group and given an appropriate name.

Questions for class discussion

1. Your scientific name is *Homo sapiens*. Which part of your name refers to your genus and which part to your species? Do you know what each of the Latin names means in English?

2. Archaeologists have discovered some fossils which are about half a million years old. They have called them *Homo habilis*, *Homo erectus*, and most recent of all *Homo neanderthalensis*. Explain as fully as you can what these names suggest about these creatures.

3. Scientific names can tell us about the relationships between organisms. Fill in the blanks in the sentences that follow.

a. The blackbird (*Turdus merula*) and song thrush (*Turdus philomelus*) belong in the same . . . .

b. The two kinds of woodlice found most frequently in this country are *Armadillidium vulgare* and *Oniscus asellus*. They are not in the same . . . . . . or . . . . . .

c. All varieties of domestic dogs from the Great Dane to the Chihuahua are named *Canis canis* so they all . . . .

Finding the name of an organism

Have you used a book to find the name of an animal or plant? It can be quite difficult, even if the book has detailed pictures.

Most identification books contain some sort of *key* to help you find your way quickly to the correct picture or description.

The simplest type of identification key is called a *spider key* and it is suitable for identifying the major groups to which organisms belong. It gets its name from its shape – an example is shown in Fig 3.7.

Individual species are usually identified using a *numbered key*. There are several examples in the Appendix of this book. They consist of a series of clues, each of which has

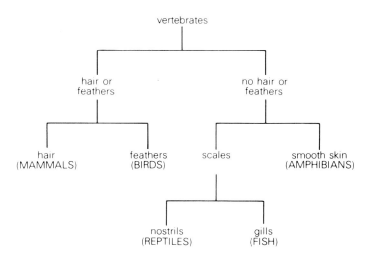

vertebrates

hair or feathers — no hair or feathers

hair (MAMMALS) — feathers (BIRDS)

scales — smooth skin (AMPHIBIANS)

nostrils (REPTILES) — gills (FISH)

Fig 3.7 Spider key to identify the groups of vertebrates

two alternative answers. The answers lead to a further clue or to the name of the species.

Using a numbered key

With practice, you should be able to use the large keys found in flower and animal books, but here is a small, simple key to start with. Use it to place the eight vertebrates in Fig 3.8 into their correct groups.

1. Examine picture (a) carefully.

2. Decide which of the alternatives in Clue 1 of the key fits this animal. You will see that it has no hair or feathers, so you should go on to Clue 3.

3. Go through all the clues until you come to the name of the group to which the animal belongs. Of course it is a crocodile and it is a reptile.

4. Write down the number of each clue you have used and the name of the group like this:

Animal A: Clues 1 $\longrightarrow$ 3 $\longrightarrow$ 4: Group REPTILE

5. Do the same for animals (b) to (h).

Fig 3.8 Put these vertebrates into groups using the key

Key to vertebrates in Fig 3.8

Clue 1 Vertebrates with hair or feathers . . Go to clue 2
Vertebrates without hair or feathers Go to clue 3
Clue 2 Hair or whiskers on face and body . MAMMALS
Feathers cover most of the body . . BIRDS
Clue 3 Body covered in scales . . . . Go to Clue 4
Body without scales; wet, smooth
skin . . . . AMPHIBIANS
Clue 4 Nostrils and lungs; lay eggs on
land . . . . REPTILES
Gills and fins; live in water . . . . . FISH

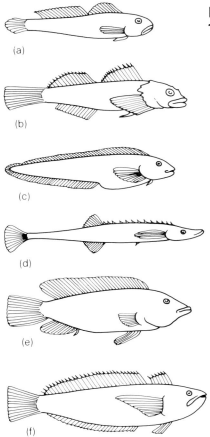

(a)

(b)

(c)

(d)

(e)

(f)

Fig 3.9 Identify these fish

Homework assignments

1. The key below identifies eight fish. Six of them are drawn in Fig 3.9. Which scientific name refers to which fish?

1. The fins or body have spines along the top (dorsal) surface . . 2
   No obvious dorsal spines 3

2. Spines on the fins 4
   Separate spines on the body but not on the fins 5

3. Single dorsal fin 6
   Two dorsal fins . . . . GOBIUS

4. Dorsal fins similar in size . . . . COTTUS
   The front dorsal fin much smaller than the rear one . . . . TRACHINUS

5. Snout upturned, body slender . . SPINACHIA
   Snout not upturned, body not slender . . . . GASTEROSTEUS

6. Distinct tail fin 7
   Dorsal fin merged with the tail fin . . . . ZOARCES

7. Elongated body; tail fin with central notch . . . . AMMODYTES
   Body not noticeably elongated; tail fin smooth . . . . LABRUS

2. Here is a key for grouping animals. Study it carefully and then answer the questions that follow it.

1. External nostrils present 2
   No external nostrils 6

2. Air-breathing 3
   Breathes in water using gills . . . FISH

3. Fur or hair absent 4
   Fur or hair present . . . . MAMMAL

4. Feathers present . . . . BIRD
   No feathers 5

5. Scales present . . . . REPTILE
Has smooth skin and no scales . . AMPHIBIAN

6. External skeleton present 7
Has no external skeleton 11

7. Body segmented (arranged in rings) 8
Body not segmented 10

8. Body divided into 3 parts, with 3 pairs of legs . . . . INSECT
Body divided into 2 parts 9

9. Four pairs of jointed legs . . . . SPIDER
Five or more pairs of jointed legs . CRUSTACEAN

10. Outer shell made of chalky substance . . . . MOLLUSC
Spiny-skinned animals . . . . . . ECHINODERM

11. Body segmented (arranged in rings) . . . . WORM
Body not segmented . . . . LOWER INVERTE-BRATES

a. To which animal does the following refer: It has *no* external nostrils; it has an external skeleton to which seven pairs of jointed legs are attached; its segmented body is divided into two parts?
b. Name *two* types of animal (from the key) that have no external nostrils, but have external skeletons that are not segmented.
c. Use the key to find *two* items of information that apply to *both* reptiles and mammals.

The animal kingdom

The vertebrates shown in Fig 3.8 belong to one of the eight main groups (or *phyla*) in the animal kingdom. All vertebrates have a backbone running down their bodies. The other six groups are all *invertebrates* (without a backbone). They can be recognised by the features shown in Fig 3.10.

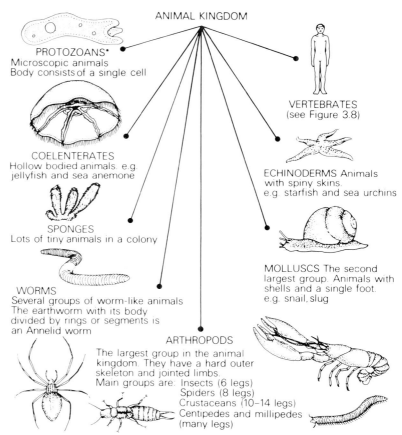

ANIMAL KINGDOM

PROTOZOANS*
Microscopic animals
Body consists of a single cell

VERTEBRATES
(see Figure 3.8)

COELENTERATES
Hollow bodied animals. e.g.
jellyfish and sea anemone

ECHINODERMS Animals
with spiny skins.
e.g. starfish and sea urchins

SPONGES
Lots of tiny animals in a colony

MOLLUSCS The second
largest group. Animals with
shells and a single foot.
e.g. snail, slug

WORMS
Several groups of worm-like animals
The earthworm with its body
divided by rings or segments is
an Annelid worm

ARTHROPODS
The largest group in the animal
kingdom. They have a hard outer
skeleton and jointed limbs.
Main groups are: Insects (6 legs)
Spiders (8 legs)
Crustaceans (10–14 legs)
Centipedes and millipedes
(many legs)

*In some classifications PROTOZOANS are put in a kingdom of their own.

Fig 3.10 The main groups in the
animal kingdom

The plant kingdom
The plants which are most familiar to us are those which
bear cones (the *conifers*) or flowers (the *flowering plants*).
However there are five other important groups of plants
shown in Fig 3.11.

Experiment 3.3 Making a collection
Making a collection of living organisms is fun. It does not
have to be large and you need not go far afield. You might
simply set some pitfall traps (see page 169) or use a butterfly

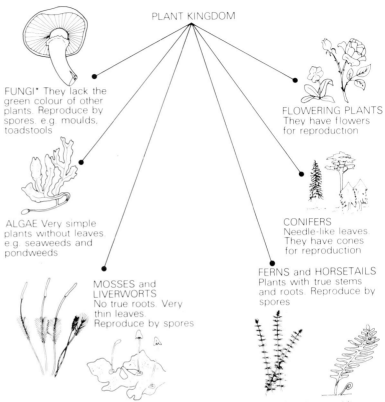

PLANT KINGDOM

FUNGI* They lack the green colour of other plants. Reproduce by spores. e.g. moulds, toadstools

FLOWERING PLANTS They have flowers for reproduction

ALGAE Very simple plants without leaves. e.g. seaweeds and pondweeds

CONIFERS Needle-like leaves. They have cones for reproduction

MOSSES and LIVERWORTS No true roots. Very thin leaves. Reproduce by spores

FERNS and HORSETAILS Plants with true stems and roots. Reproduce by spores

Fig 3.11 The main groups in the plant kingdom

*In some classifications, FUNGI are placed in a kingdom of their own. *Viruses* and *Bacteria* are also put into separate kingdoms.

net or collect fungi in the autumn. However there are a few rules which you should observe when collecting.

1. Take notes, drawings or photographs where possible. Never take more specimens than you need.

2. Leave flowers for others to enjoy, and birds' eggs to hatch and develop.

3. Consider other people's property. Always ask permission if in doubt. It will usually be granted readily.

4. Do not disturb animals or plants if you can help it. Even visiting a birds' nest may lead to the parents deserting it, so take care.

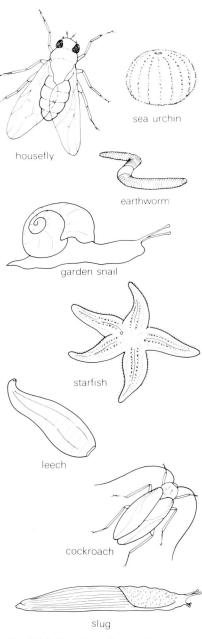

housefly

sea urchin

earthworm

garden snail

starfish

leech

cockroach

slug

Fig 3.12 Put these invertebrates into groups using the key

There are many books which give helpful advice about collecting, identifying and mounting specimens. Some are listed in the Appendix. Do not worry if you cannot give a name to everything you find; just identify it as far as you can. Present your collection in an attractive and informative way so that the rest of the class can enjoy looking at it.

Homework assignments

1. Complete the following sentences by selecting the best word or phrase.
a. Man is classified as a mammal because
   he has hair on his body
   he is warm blooded
   he has an internal skeleton
b. The molluscs include
   jellyfish
   snails
   earthworms
c. Fungi, mosses and ferns reproduce by
   spores
   seeds
   larvae

2. Look carefully at the drawings of the eight invertebrates shown in Fig 3.12. Copy down the numbered key shown below, filling in each of the blanks where indicated.

1. Body in 5 parts, covered in spines 2
   Body not arranged in 5 parts or covered in spines. 3

2. . . . . STARFISH
   . . . . SEA URCHIN

3. Three pairs of legs or limbs 4
   No legs or limbs; body mainly smooth 5

4. . . . . HOUSEFLY
   . . . . COCKROACH

39

5. Antennae or feelers on top of
   head. 6
   No antennae or feelers visible. . . . 7

6. . . . . LEECH
   . . . . EARTHWORM

7. SLUG
   . . . . GARDEN SNAIL

3. Suppose a spacecraft returns from a mission to a newly discovered moon of Saturn with several specimens of living organisms. Describe how you would go about the task of classifying and naming them.

Background reading

Fossils

The vast majority of living things leave no trace of their existence after their passing. Their flesh decays, their shells and bones become scattered and turn to powder. Very occasionally, one or two individuals out of a population of many thousands have a different fate. A reptile becomes stuck in a swamp and dies. Its body rots but its bones settle in the mud. Dead vegetation drifts to the bottom and covers them. As the centuries pass and more vegetation accumulates, the deposit turns to peat. Changes in sea level may cause the swamp to be flooded and layers of sand to be deposited on top of the peat. Over great periods of time, the peat is compressed and turns to coal. The reptile's bones still remain within it. The great pressure of the overlying sediments and the mineral-rich solutions that circulate through them cause chemical changes in the calcium phosphate of the bones. Eventually they are turned to stone.

The most suitable places for fossilisation are in seas and lakes where sedimentary deposits such as sandstones and limestones are slowly accumulating. On land, where for the most part rocks are not built up by deposition but broken down by erosion, deposits, such as sand dunes, are only very rarely created and preserved. In consequence, the only land-living creatures likely to be fossilised are those that happen to fall into water. Since this is an exceptional fate for most of them, we are never likely to know from fossil evidence anything approaching the complete range of land creatures that has existed in the past. Water-living animals such as fish, molluscs, sea urchins and corals, are much more promising candidates for preservation. Even so, very few of these perished in the exact physical and chemical conditions necessary for fossilisation. Of those that did, only a tiny proportion happen to lie in the rocks that outcrop on the surface of the ground today; and of these few, most will be eroded away and destroyed before they are discovered by fossil hunters. The astonishment is that, in the face of these adverse odds, the fossils that have been collected are so numerous and the record they provide so detailed and coherent.

(From *Life on Earth* by David Attenborough, published by Collins/BBC.)

Fig 3.13 A fossil ammonite; this was once a water-living animal

Questions

1. Write down the meaning of the words which are underlined in the text.

2. What prevents the bones of the reptile from decaying?

3. Explain in your own words why water-dwelling animals are more frequently found in the fossil record than land-dwelling animals.

4. Suggest one other way in which animals or plants can be preserved apart from fossilisation.

5. Find out, and report on, *one* of the following:
a. What fossils tell us about the first land animals
b. The fossil fish, Coelocanth
c. The fossil record
d. The rise and fall of the dinosaurs

Summary

1. Dividing things into groups is a process called *classification*.

2. The animal and plant kingdoms are divided into main groups called *phyla*.

3. Each phylum is divided into sub-groups which contain many *genera*.

4. Within each *genus* there are one or more individual *species*.

5. Each species is identified by two Latin names. The *first* refers to its *genus* and the *second* refers to the *species* itself.

6. *Keys* are used to find the names of species or the group to which they belong.

7. There are 950 000 named species in the animal kingdom and about 340 000 named species in the plant kingdom.

Chapter 4 # Microscopes and cells

To find out about the detailed structure of an animal or plant, we have to look at small pieces of its body under the *microscope*.

In this chapter we shall start by learning about the microscope, and then we shall use it to study the structure of some animals and plants.

Preparing specimens for viewing under the microscope

Usually a specimen that is to be viewed under the microscope is first placed in a drop of fluid on a rectangular piece of glass called a *slide*. It is then covered with another, much thinner, piece of glass called a *coverslip* (Fig 4.1). The coverslip flattens the specimen and prevents the fluid evaporating from underneath. When a specimen is prepared in this way it is said to have been *mounted*.

The fluid in which the specimen is mounted may be pure water, or it may be some kind of stain. Transparent specimens are mounted in a stain because it helps them to show

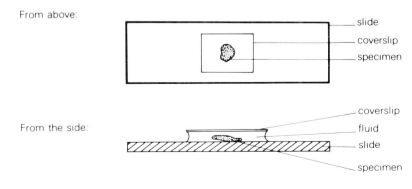

From above:

slide
coverslip
specimen

From the side:

coverslip
fluid
slide
specimen

Fig 4.1 A specimen mounted on a slide ready for viewing under the microscope

up. The disadvantage of using a stain is that it usually kills the specimen.

Sometimes the specimen is mounted in a transparent cement which sets hard. Specimens mounted in this way are called permanent preparations, and can be examined under the microscope years later.

The microscope

A microscope can magnify objects much more than a simple hand lens can. It has two lenses, one above the other, and it has special devices for lighting and focusing the specimen.

A typical school microscope is shown in Fig 4.2. Not all microscopes are like this one. For example, some models have a built-in light under the stage.

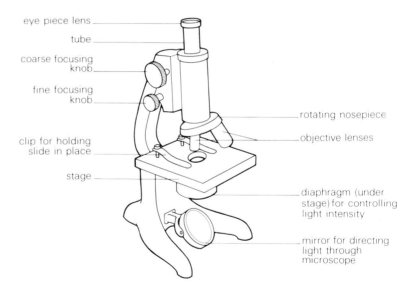

eye piece lens
tube
coarse focusing knob
fine focusing knob
clip for holding slide in place
stage
rotating nosepiece
objective lenses
diaphragm (under stage) for controlling light intensity
mirror for directing light through microscope

Fig 4.2 A microscope of the kind that you may have in your school laboratory

Experiment 4.1   Learning to use the microscope

Study your microscope carefully and find its various parts. Use Fig 4.2 to help you. Make sure you know the name of

Fig 4.3 Setting up a microscope

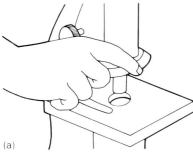

(a)

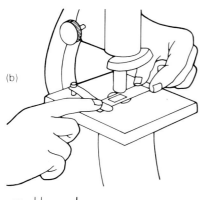

(b)

each part, and understand what it is for, before you start using the microscope.

Your teacher will give you a specimen which has been mounted on a slide. To set it up under the microscope, do this:

1. Rotate the nosepiece of your microscope so that the *small* objective lens is immediately above the centre of the stage: the nosepiece should click into position (Fig 4.3(a)).

2. Place the slide on the stage. Arrange it so that the specimen is in the centre of the hole in the stage. When you have positioned the slide, fix it in place with the clip (Fig 4.3(b)).

3. Place a lamp in front of the microscope, and set the angle of the mirror so that the light is directed up through the microscope (Fig 4.3(c)).

4. Look down the microscope through the eyepiece. The round, bright area that you can see is called the *field of view*. Adjust the diaphragm so that the field of view is not too bright (Fig 4.3(d)).

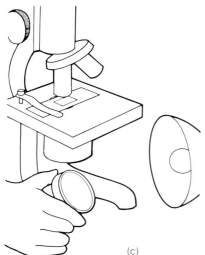

(c)

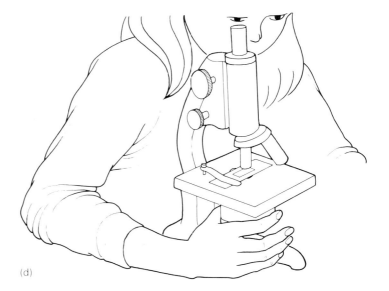

(d)

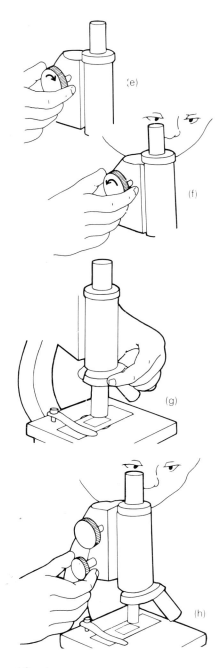

5. Look at the microscope from the side. Turn the coarse focusing knob in the direction of the arrow in Fig 4.3(e). We call this 'racking down': it moves the tube downwards. Continue turning the knob until the tip of the objective lens is about 10 mm from the coverslip.

6. Now look down the microscope again. Slowly turn the coarse focusing knob in the other direction. We call this racking up: it moves the tube upwards (Fig 4.3(f)). The specimen on the slide should soon come into view. Now use the fine focusing knob to focus the specimen as sharply as possible.

7. If necessary, re-adjust the diaphragm and the angle of the mirror so that the specimen is correctly lit. *You will get a much clearer picture if you do not have too much light coming through the microscope.*

   You are now looking at the specimen under *low power*, i.e. at a low magnification. To look at it under *high power*, i.e. at a greater magnification, do this:

8. Rotate the nosepiece so that the *large* objective lens is immediately above the specimen (Fig 4.3(g)). The nosepiece should click into position as before. You will find that the lens is very close to the coverslip.

9. Look down the microscope. The specimen should be in focus. If it is not in focus, a very slight movement of the fine adjustment knob should bring it into focus (Fig 4.3(h)). If, having done this, you still cannot see the specimen, ask your teacher to help you.

10. If necessary, increase the lighting by opening the diaphragm or altering the angle of the mirror. You are now looking at the specimen under high power. You will find that it appears much larger than under low power.

Important rules for using the microscope

1. Always treat the microscope with great care: it is an expensive instrument.

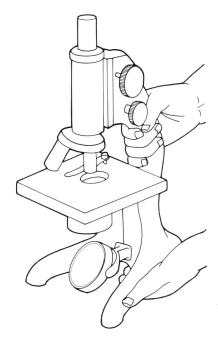

Fig 4.4 The right way to carry a microscope

2. Carry it with *both* hands as shown in Fig 4.4, and put it down *gently* on the laboratory bench.

3. Cover the microscope when you are not using it.

4. Make sure the lenses never get scratched or damaged. Clean them with lens tissue only, *never* with a handkerchief.

5. Never rack downwards with the coarse adjustment knob while you are looking down the microscope.

6. When changing slides make sure that the tip of the objective lens is well clear of the coverslip.

Questions for class discussion

1. When mounting a specimen it is important not to put too much fluid on the slide. What would happen if you were to put on too much fluid, and what might the consequences be?

2. Why should you never rack downwards with the coarse focusing knob while you are looking down the microscope?

3. Suggest two reasons why it is a bad idea to have too much light coming through your microscope.

The magnifying power of the microscope

The magnification which can be achieved by your microscope is the magnifying power of the eyepiece multiplied by the magnifying power of the objective lens. The magnifying powers of the eyepiece and the objectives are usually engraved on them. Find these figures and then work out the magnification of your microscope on low and high powers.

Copy Table 4.1 into your notebook and fill in the first three columns.

Table 4.1

	Magnifying power		Total magnifi-cation	Diameter of field of view
	Eyepiece lens	Objective lens		
Low power				
High power				

Experiment 4.2   Estimating the size of an object under the microscope

1. Place a transparent ruler, or graduated slide with a milli-metre scale, on the stage of your microscope. Using the low power objective lens, focus on to the lines on the scale.

2. Count how many millimetre divisions fit across the low power field of view. What is the diameter of the low power field of view in millimetres?

3. Now rotate the nosepiece so that the high power objec-tive lens is immediately above the scale. What is the approximate diameter of the high power field of view to the nearest tenth of a millimetre?

4. Fill in the fourth column of your table of magnifications.

5. Remove a hair from your head, place it on a slide and look at it under the low power. Estimate its width in millimetres.

Homework assignments
1.
a. If the magnifying power of the eyepiece of a microscope is ×10, and that of the high power objective is ×40, what will be the total magnification which this micro-scope can achieve?

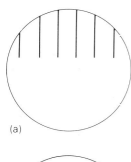

(a)

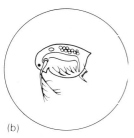

(b)

Fig 4.5

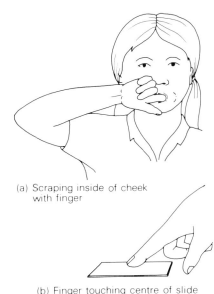

(a) Scraping inside of cheek with finger

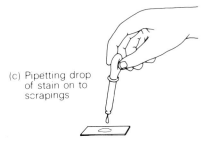

(b) Finger touching centre of slide

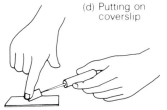

(c) Pipetting drop of stain on to scrapings

(d) Putting on coverslip

Fig 4.6 The procedure for looking at cheek cells under the microscope

b. Suppose that a microscope has two objective lenses ($\times 10$ and $\times 40$) and two alternative eyepieces ($\times 6$ and $\times 10$). Write down the four different magnifications which this microscope can achieve. In each case show how you arrive at your answer.

2. A student tries to work out the diameter of the field of view of her microscope using a transparent ruler with a millimetre scale. Fig 4.5(a) shows what she sees when she looks at the ruler under the microscope. She then looks at a little animal under the microscope at the same magnification, and its appearance is shown in Fig 4.5(b). Work out as accurately as possible:

a. the diameter of the field of view
b. the actual length of the animal.

3. A schoolboy attempts to look at a specimen under the microscope, but he cannot see anything at all. Make a list of as many reasons as you can think of for his lack of success.

Cells

A house is made of bricks. In the same way, the bodies of animals and plants are made of *cells*. A cell is really more like a little box than a brick, and it contains some important structures which we will look at presently.

A fully-grown human is made of about one hundred million million cells. If the body is made of so many cells, each individual cell must be very small. In fact they can only be seen with a microscope.

Experiment 4.3 Looking at cheek cells

1. Wash your hands, then gently scrape the inside of your cheek with a finger (Fig 4.6(a)).

2. Dab the tip of your finger on to the middle of a clean microscope slide (Fig 4.6(b)).

3. With a pipette, add a drop of 0.5 per cent methylene blue solution to the scrapings on the slide (Fig 4.6(c)). This will stain any cells present and help to show them up.

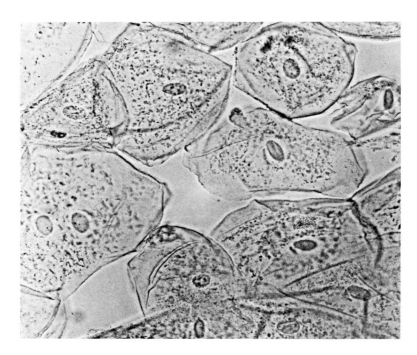

Fig 4.7 Human cheek cells as they appear under the microscope, high power

4. Carefully cover the scrapings with a coverslip. Support one side of the coverslip with a finger, and lower the other side with a needle (Fig 4.6(d)). If you lower the coverslip gently, you will not get any air bubbles underneath.

5. Examine your slide under the low power of the microscope. Find some of the scrapings. Can you see lots of semi-transparent objects each with a blob inside? Each semi-transparent object is a cell. The blob is oval in shape, rather like a rugger ball.

6. Now look at one of the cells under high power (Fig 4.7). Make a careful drawing of it in pencil. Do not use colour and do not label it for the moment.

Questions for class discussion

1. Some of your cheek cells may look crumpled. What does this tell us about them?

2. You may have noticed that some of the cells, instead of being on their own, are in small groups and fit together rather like 'crazy paving'. What job do you think these cells carry out inside your mouth?

3. To do this experiment you had to remove some of the cells from inside your mouth. Do you think this may have harmed you? If not, why not?

The structure of a typical animal cell

The cheek cell you have just looked at is a typical animal cell. The main parts of a typical animal cell are shown in Fig 4.8.

The *cell membrane* forms a very thin covering to the cell. It controls what enters and leaves the cell: it allows oxygen and dissolved food substances to get in and waste substances to get out.

The *cytoplasm* makes up most of the inside of the cell. In the cytoplasm, many important chemical reactions take place. Some of these reactions release energy which helps to keep the cell alive. If you look at the cytoplasm carefully under the microscope, you will see that it contains numerous tiny dots. We call these *granules*: some of them consist of stored food.

The *nucleus* is the dark blob in the centre of the cell. It is the control centre of the cell, regulating everything that goes on inside it. The nucleus contains special chemical structures called *genes* which are passed from parents to offspring. The genes determine the person's features such as the colour of his or her eyes.

The nucleus and cytoplasm together make up the *protoplasm*.

Now that you know the parts of a typical animal cell, label your drawing of the cheek cell.

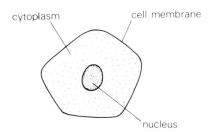
cytoplasm   cell membrane
nucleus

Fig 4.8 A typical animal cell

Experiment 4.4   Estimating the size of a cheek cell

To do this experiment you will need to know the width (that

is, the diameter) of the high power field of view of your microscope. This is explained in Experiment 4.2. (p. 48). You can use your slide from Experiment 4.3 for this experiment, or you can make a new slide.

1. Locate a cheek cell which has a regular shape, and look at it under high power.

2. As accurately as possible, estimate how many cells of this size would fit side-by-side across the diameter of the field of view.

3. Knowing the diameter of the field of view, work out the width in millimetres of the cheek cell.

4. Look back at your drawing of a cheek cell and with a ruler, measure its width in millimetres. How many times wider is your drawing than the actual cell? This figure is the magnification of your drawing. Write the figure underneath your drawing with a times sign (×) in front of it. You have now given your drawing a *scale*.

Expressing the size of a cell
A typical animal cell is approximately one fiftieth of a millimetre wide. It is awkward to express such a short distance as a fraction of a millimetre, and so a smaller unit is used. This is called a *micrometre*. A micrometre is one thousandth of a millimetre. If a cell is one fiftieth of a millimetre wide, what is its width in micrometres? The answer, of course, is 20 micrometres.

The abbreviation for a millimetre is mm. The abbreviation for a micrometre is μm.

A typical cell is $\frac{1}{50}$ mm wide, which is 20 μm.

Look back to your drawing of the cheek cell. What is the width of the actual cell in millimetres? Convert this to micrometres and write it under your drawing.

Experiment 4.5   Looking at moss cells
You will need a piece of moss which has been carefully pulled away from a path or wall.

1. Take off one of the little leaves. Place the leaf on a slide and add a drop of water to it. Cover it with a coverslip.

2. Look at the leaf under the low power of the microscope. Can you see that it is composed of numerous cells which are arranged rather like bricks in a wall?

3. Examine one cell under high power. Make a drawing of it in pencil. Do not label it yet.

4. Estimate the scale of your drawing, and write this under the drawing.

5. Work out the length of the actual cell in micrometres, and write it under the drawing.

6. With the fine adjustment knob, rack up half a turn and then down half a turn. How does the appearance of the cell change as you rack up and down? What does this tell you about the three-dimensional shape of the cell?

Questions for class discussion
1. Compare your drawings of the cheek cell and the moss cell:
a. Which one is larger?
b. How do they differ in shape?
c. What structures could you see in the moss cell which were not visible in the cheek cell?

2. You were probably unable to see a nucleus in the moss cell. Suggest reasons for this.

The structure of a typical plant cell

A typical plant cell has the cell membrane, cytoplasm and nucleus typical of animal cells. However, it has three extra structures which are found *only* in plant cells. These are chloroplasts, a vacuole and a cell wall (Fig 4.9).

The *chloroplasts* contain a green substance called *chlorophyll*. This is what makes leaves and stems look green. Chlorophyll is needed for the process by which plants make food. We call this process *photosynthesis*.

The *vacuole* is a large space in the centre of the cell. It is filled with a watery fluid containing sugar and salts.

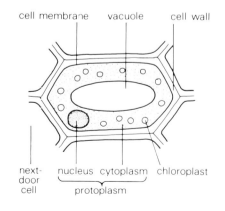

cell membrane   vacuole   cell wall

next-door cell   nucleus   cytoplasm   chloroplast

protoplasm

Fig 4.9   A typical plant cell

The *cell wall* is situated outside the cell membrane. It is much thicker and tougher than the cell membrane and is made of a substance called *cellulose*. Cellulose is important to man because paper and various other products are made from it.

Now that you know the structure of a typical plant cell, label your drawing of the moss cell.

Homework assignments

1. Copy out Table 4.2. Put a tick in the box if the structure listed on the left is present, and a cross if it is not.

Table 4.2

	Typical animal cell, for example, cheek cell	Typical plant cell, for example, moss leaf cell
Cell membrane		
Cell wall		
Chloroplasts		
Cytoplasm		
Nucleus		
Vacuole		

2. Which of the structures listed in the previous question
a. provides us with a source of paper
b. is the control centre of the cell
c. carries out photosynthesis
d. releases energy to help keep the cell alive
e. is filled with a watery fluid?

3. Fig 4.10 shows a group of cells.
a. Name the structures labelled A, B and C.
b. Are the cells from an animal or a plant?
c. Give four reasons to support your answer to b.

4. Your friend was away when you looked at cells under the microscope. Explain to her how you obtained a sample of cheek cells and prepared them for viewing under the microscope.

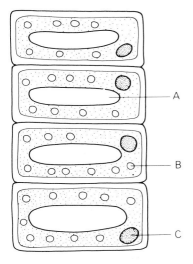

Fig 4.10

54

5. A girl observes a cell down the microscope and she estimates it to be $\frac{1}{100}$ mm wide.

a. What is the cell's width in micrometres (μm)?

b. Why is it better to express its width in micrometres rather than millimetres?

c. The girl draws the cell, and the width of her drawing is 25 mm. What is the scale of her drawing?

Different cells for different jobs

The cheek cells which you looked at earlier are just one kind of cell found in the human body. Their job is to form the lining of the inside of the mouth.

There are many other kinds of cells in the human body, and each one has a particular job to do. For example, the brain contains cells which transmit electrical messages from place to place, a muscle is made up of cells which can shorten and thereby move the animal, and blood is composed of cells which carry oxygen around the body.

The same principle applies to plants. For example, most of the cells in a leaf contain chloroplasts and carry out photosynthesis: they are responsible for feeding the plant. However, cells in the root do not have chloroplasts: they carry out the important job of taking up water and mineral salts from the soil.

How do cells multiply?

When an animal or plant grows, it gets larger. To enable the organism to increase in size, its cells divide as shown in Fig 4.11. First the nucleus splits into two and then the cell divides across the middle. This is called *cell division*. The two cells resulting from the division are called *daughter cells*. The daughter cells now expand to their full size, after which each one may divide again. The result will be a total of four cells, and if each of these then divides again there will be eight cells – and so on.

How can you tell if a cell has just divided? You can usually tell from the size of the daughter cells: they will be smaller

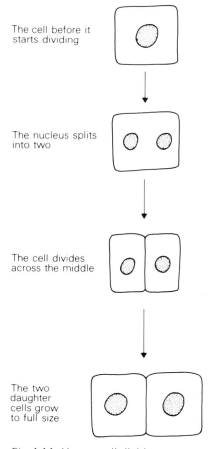

The cell before it starts dividing

The nucleus splits into two

The cell divides across the middle

The two daughter cells grow to full size

Fig 4.11 How a cell divides

than a full grown cell of the same kind. A good place to see cells that have just divided is in a simple water-weed.

Experiment 4.6 Looking at divided cells in a water-weed

The water-weed you will look at consists of long slimy threads. It is found in ponds and slow-flowing rivers where it occurs in clumps.

1. With a pair of tweezers place a few threads of the water-weed on a slide, together with a drop of water, and cover it with a coverslip.

2. Look at the threads under the microscope, low power first then high power. Notice that each thread is divided up into a series of compartments. Each compartment is a cell (Fig 4.12). The cells look green because they contain chloroplasts.

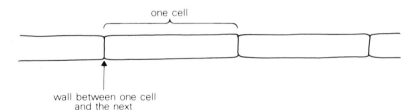

Fig 4.12 Part of one thread of a water weed

3. Compare the sizes of the cells. You may find that some are shorter than others. If you find two short cells side by side, they were probably formed by a larger cell dividing into two. This kind of cell division enables the threads to get longer; in other words it results in growth.

Experiment 4.7 Looking at small organisms under the microscope

We have seen how the microscope can be used to study cells. It can also be used to study little organisms which are too small to be seen with the naked eye or magnifying glass.

1. Obtain a jam jar of water from the bottom of a pond or ditch: the dirtier the better.

Euglena (40 μm)   Paramecium (250 μm)   Roundworm (2–5 mm)    Rotifer (1 mm)    Stylonichia (0.5 mm)

Fig 4.13 Some small pond water organisms which you might see under your microscope. The approximate lengths of the organisms are given after their names

2. Place a drop of the water on a slide and cover it with a coverslip.

3. Examine it under the microscope, low power first, then high power.

Can you see any small organisms swimming around? Fig 4.13 shows some of the organisms which you may see under the microscope.

Single-celled organisms

Some of the little organisms shown in Fig 4.13 consist of only one cell. There are many kinds of single-celled organisms and some of them live in ponds and ditches. Different kinds live inside the bodies of other organisms: they feed on the living material inside the body and may do a great deal of harm. The disease *malaria* is caused by a single-celled organism which lives in the human bloodstream. These single-celled organisms therefore matter a lot to man.

Homework assignments

1. Although chloroplasts are found in many plant cells, they are not always present. Suggest two places in a dandelion plant where you would *not* expect to find chloroplasts in the cells.

2. The end of a root is covered with delicate extensions called root hairs. Each hair is a single cell, which looks like Fig 4.14 under the microscope.

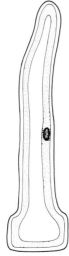

Fig 4.14 A root hair

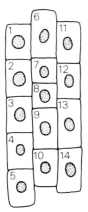

Fig 4.15 Dividing cells

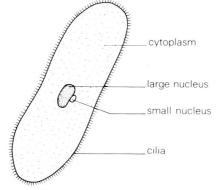

Fig 4.16 *Paramecium*

a. Describe how you would prepare the root of a young plant in order to look at its cells under the microscope.
b. Write down two ways in which this cell differs from a moss leaf cell.
c. What do you think its function might be?

3. The cells inside the tip of a root divide repeatedly; as a result, the root grows and gets longer. Fig 4.15 shows a group of such cells inside a bean root. Which two cells were formed by a recent cell division? (Give their numbers.) How do you know?

4. Fig 4.16 shows a single-celled organism called *Paramecium* which lives in ponds and ditches. It is covered with little hairs called *cilia* which beat backwards and forwards, thereby propelling the animal through the water.
a. Do you think this organism is an animal or plant? Give reasons for your answer.
b. Write down three ways in which it differs from a cheek cell.
c. The scale of the diagram is × 216. What is the actual length of this organism?

Background reading

Cells were discovered in the 17th century by an English scientist, Robert Hooke. In the following passage, Hooke explains how he made his discovery. Beneath the passage is a reproduction of Hooke's original drawing of cork cells.

'I took a good clear piece of cork, and with a Penknife sharpen'd as keen as a Razor, I cut a piece of it off, and thereby left the surface of it exceedingly smooth, then examining it very diligently with a Microscope methought I could perceive it to appear a little porous; but I could not so plainly distinguish them, as to be sure that they were pores . . . . . I with the same sharp Penknife, cut off from the former smooth surface an exceeding thin piece of it, and placing it on a black object Plate, because it was itself a white body, and casting the light on it with a deep plano-convex Glass,

I could exceedingly plainly perceive it to be all perforated and porous, much like a Honeycomb, but that the pores of it were not regular .... these pores, or cells, were not very deep, but consisted of a great many little Boxes .... Nor is this kind of texture peculiar to Cork onely; for upon examination with my Microscope, I have found that the pith of an Elder, or almost any other Tree, the inner pulp or pith of the Cany hollow stalks of several other Vegetables: as of Fennel, Carrets, ... Teasels, Fearn .... etc. have much such a kind of Schematisme, as I have lately shewn that of Cork.'

(From *Micrographia* by Robert Hooke, published by the Wellcome Trustees.)

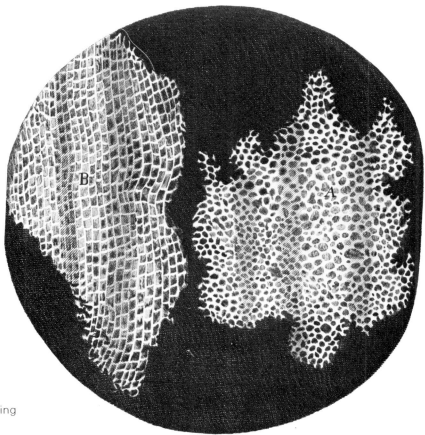

Fig 4.17 Robert Hooke's drawing of cork cells

Questions

1. Why do you think Hooke chose cork for examination under the microscope?

2. How do the cork cells in his drawing differ from the cheek cells shown in Fig 4.7? Why do you think they are different?

3. Hooke described his cells as 'pores' and as 'boxes'. Which word is the better one, and why?

Summary

1. We use a *microscope* for looking at organisms, or parts of organisms, which are too small to be seen with the unaided eye or hand lens.

2. When observing a specimen under the microscope, it is important to know the *magnification* and to record the *scale*.

3. A typical animal cell contains *cytoplasm* and a *nucleus* and is surrounded by a *cell membrane*. It is approximately $\frac{1}{50}$ mm wide ($20 \, \mu$m).

4. A typical plant cell has cytoplasm, a nucleus and cell membrane, plus *chloroplasts*, a *vacuole* and *cellulose cell wall*.

5. A complex organism such as the human contains millions of cells. Different kinds of cells carry out different jobs.

6. Cells multiply by splitting into two (*cell division*).

7. Many organisms exist which can be seen only under a microscope. Some of these organisms consist of only one cell.

Chapter 5 # Asexual reproduction

Reproduction is the process by which an organism produces new individuals (offspring). Cell division, which we studied in the last chapter, is the basis of reproduction.

In humans and many other organisms reproduction involves two individuals, the male and the female. We call this sexual reproduction. However, some organisms can reproduce on their own without another individual. We call this asexual reproduction. In this chapter, we shall look at some of the methods by which organisms reproduce asexually.

Splitting in two

Amoeba is a single-celled organism which lives in ponds and puddles (see p. 37). It reproduces by splitting in two. We call this *binary fission* (*fission* means 'splitting' and *binary* means 'two'). The process is just like cell division described in the previous chapter (p. 55): first the nucleus divides into two, and then the cell splits across the middle as shown in Fig 5.1. The two new amoebae then grow, and after a day or so each of them may split again.

Questions for class discussion
1. When it is warm and plenty of food is available, *Amoeba* may split once every 24 hours. Dividing at this rate, how many amoebae could be formed from a single *Amoeba* after seven days?

2. Bacteria also multiply by cell division but they do so much faster than *Amoeba*: in good conditions division may occur once every 20 minutes. At this rate, how many cells would be formed from a single bacterial cell after 24 hours?

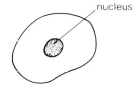
nucleus
The cell rounds off

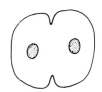
The nucleus splits into two

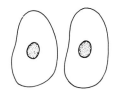
The cell has divided into two daughter cells

Fig 5.1 *Amoeba* reproduces asexually by dividing into two

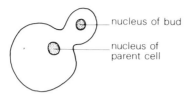

(a) Single bud

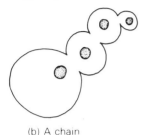

(b) A chain

Fig 5.2 Yeast cells budding

Budding

Yeast is another single-celled organism, in this case a fungus. Wild yeast grows on the surface of fruit where it feeds on sugar. Yeast is also cultivated by man because it is used in making alcoholic drinks and bread.

Yeast reproduces by *budding*. The cell sends out a small outgrowth which gets larger and eventually breaks away from the parent cell (Fig 5.2(a)). Meanwhile the nucleus divides into two. One of the two resulting nuclei stays in the parent cell, and the other one moves into the bud.

Sometimes the new cell starts budding before it has broken away from the parent cell, thus giving rise to a chain of cells as shown in Fig 5.2(b).

Experiment 5.1   Observing yeast budding

Yeast can be obtained from a baker's or chemist's shop as a putty-like substance. It consists of millions of yeast cells stuck together.

1. Pour a ten per cent solution of glucose into a test-tube until it is three-quarters full.

2. Add a pinch of yeast to the test-tube and shake well. The cells will separate from the solid yeast and form a suspension in the glucose solution. Leave your test-tube in a warm place overnight.

3. Next day swirl the flask, then transfer a drop of the yeast suspension on to a slide. Put on a coverslip. Examine the drop under the microscope, low power first, then high power. Can you see any yeast cells with buds? Make a drawing of a budding yeast cell.

Questions for class discussion

If you sniff the contents of your test-tube you will find that it has a distinctive smell.

1. What substance is giving the yeast suspension this smell?

2. How, and why, was this substance formed?

3. Yeast is used in baking bread because it makes the dough rise. How do you think the yeast does this?

Spores

The mushroom is also a fungus, but a very different one from yeast. The mushroom, like many other fungi, reproduces by means of *spores*. A spore is a tiny round cell enclosed within a thick protective wall. The wall enables the spore to withstand unfavourable conditions such as frost and drought.

The spore is like a tiny speck of dust and is so light that it can float through the air. If it lands in a suitable place it breaks open and gives rise to a new fungus.

Experiment 5.2   To see the spores produced by a mushroom or toadstool

1. Obtain a fully-grown mushroom or toadstool with a large flat cap. Look at the underside of the cap. Can you see a series of delicate brown membranes fanning out from the stalk, like the spokes of a bicycle wheel? The spores are formed on these membranes (Fig 5.3).

2. Cut off the stalk as high up as possible. Place the cap, lower surface downwards, on a sheet of white paper. Cover it with a dish, and leave it for two or three days.

3. After two or three days remove the dish and carefully lift up the cap. Can you see any spores on the sheet of paper? Approximately how many are there? What you have produced is a *spore print*.

4. With a dry paintbrush, transfer a few spores to a slide. Add a drop of water and put on a coverslip. Examine the spores under the low power of the microscope. Make a rough estimate of the width of one of the spores (see p. 51).

5. Make your spore print permanent by spraying it lightly with an aerosol varnish. You can then stick it in your notebook.

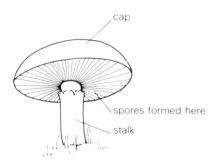

Fig 5.3 The spore-forming body of a mushroom or toadstool

cap

spores formed here

stalk

Question for class discussion

Spores are formed by these three main groups of organisms: fungi, mosses and ferns. As well as providing a method of reproduction, spores enable the species to spread rapidly over a wide area. How do they achieve this?

Tubers

The potato plant forms underground structures called *tubers* which rest in the soil during the winter and produce new plants the next year.

The tuber is the familiar 'potato' which people eat. It is full of stored food which is made by the leaves of the parent plant during the summer. In the autumn, the leaves and stem of the parent plant die, but the tubers remain in the soil until the following spring. Then each one may sprout into a new potato plant.

Tubers enable potato plants to reproduce asexually. They also enable the species to survive the winter so that potato plants can grow up again year after year.

Experiment 5.3 Examining a potato tuber

1. You may have heard people talk about the 'eyes' of a potato. They are shown in Fig 5.4. Can you see any 'eyes' on your potato tuber? Each one is a very small bud which is capable of sprouting into a new potato plant.

   You will also see little black specks dotted about over the skin of the potato. These are little holes called *lenticels* which allow air to get through the skin so that the tuber can breathe.

2. Cut a potato tuber in two. With a knife, scrape away a little of the white pulp and put it on a slide. Add a drop of weak iodine solution to the pulp, and cover it with a coverslip.

3. Look at the potato pulp under the microscope. Do you see a lot of dark blue, egg-shaped objects? These are *starch grains*.

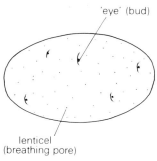

'eye' (bud)

lenticel
(breathing pore)

Fig 5.4 A potato tuber

4. Keep another potato in the window of your laboratory. You may find that eventually one or more leafy shoots start growing out of it.

Questions for class discussion

1. Where did the starch in the potato tuber come from? What will happen to it eventually?

2. What disadvantages has the method of reproduction used by potato tubers compared with reproduction by spores?

3. As well as providing a means of reproduction, potato tubers are useful to the plant for another reason. What is this reason?

Experiment 5.4   Growing potatoes

You will need a 'seed' potato which is just beginning to sprout.

1. Remove all the shoots except one. Rub them off with your thumb or finger (Fig 5.5).

2. Plant the tuber in a large pot of well watered soil mixed with compost. The tuber should be covered by about 10 cm of soil, and the shoot should be pointing upwards. Put the pot in a well lit place.

3. When the young plant is about 15 cm high, spread some compost on the soil so that it covers the buds at the bottom of the stem (Fig 5.6). These buds will then produce underground branches which will swell up to form young tubers ('new potatoes').

4. Loosen the soil every now and again, and water it regularly. Add more compost every few weeks so as to keep the lowest buds well covered.

5. After about two months dig the plant up carefully and wash the soil off the roots and tubers. How many new tubers has the potato plant produced? Can you see the remains of the old tuber? What does it look like now?

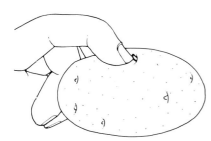
Fig 5.5 Rubbing unwanted shoots off a potato tuber

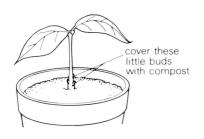
cover these little buds with compost

Fig 5.6 How to encourage a potato plant to form tubers

65

Questions for class discussion

1. Why should you remove most of the shoots from the tuber before you plant it?

2. Why is it important to loosen the soil every now and again?

3. When potatoes are grown out of doors, the tubers should be planted at least 30 cm apart. Why?

Bulbs

Daffodils, onions, hyacinths and many other plants can re-produce and survive the winter as *bulbs*. A bulb is like a tuber in that it contains a store of food. However, its structure is more complicated. It can sprout into a new plant, and it can also give rise to new bulbs.

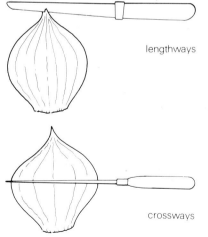

lengthways

crossways

Fig 5.7 How to slice a bulb to see its internal structure

Experiment 5.5　Examining a bulb

You will need two bulbs: onions are suitable.

1. Slice one of the bulbs lengthways, and the other one crossways (Fig 5.7). How does the inside of the bulb differ from that of the potato? What part of the plant do you think the bulb is formed from?

2. In the centre of the bulb there is a small bud which will sprout into a new onion plant. Can you see the bud in your sliced bulbs?

3. A bulb is capable of producing new bulbs. If you look at a daffodil bulb you can sometimes see small 'daughter' bulbs attached to the side of the parent bulb.

4. Fill a milk bottle with water and place a daffodil bulb on top of it so that the roots dangle in the water. Put it in a warm, well lit place. Watch it at intervals during the next few weeks and describe what happens.

Question for class discussion

When you sliced your bulb in half did you notice that it consists of a series of 'rings'? These are actually leaves. In what ways do they differ from normal leaves? Explain the reasons for the differences.

Experiment 5.6   Growing daffodils from bulbs

You will need a 'prepared' bulb and a pot of good, well watered soil mixed with compost.

1. Push the bulb into the soil. The top of the bulb should be about 4 cm below the surface of the soil.

2. Place the pot in a well lit place, and observe it every day for the next few weeks. Water the soil regularly. Record what happens in your notebook.

3. After about two months dig the plant up carefully and wash the soil off the bulb. What does the bulb look like now?

Questions for class discussion

1. Covering the bulb with soil when you plant it ensures that it will be in the dark. This causes roots to grow down into the soil before the shoot starts developing. Why is this desirable?

2. Many plants which flower in the spring have bulbs. Can you suggest why?

3. A bulb will generally produce a flowering shoot only if it is first subjected to a period of cold.
a. Why do you think this is?
b. Describe an experiment which you could do to find out how low the temperature has to be for the bulb to produce a shoot later.

Cuttings

If a side branch of a plant is cut off and stuck into some soil

(a)

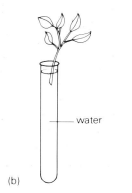

water

(b)

Fig 5.8 How to take a cutting of a Busy Lizzie plant

Fig 5.9 How to plant a Busy Lizzie cutting

or compost, roots may grow out of it so that it becomes a new plant. This is called a *cutting*.

Gardeners often take cuttings of plants which they particularly like.

Experiment 5.7 Taking cuttings of Busy Lizzie

Busy Lizzie is a plant from which it is particularly easy to take cuttings. You will need a test-tube in a rack, and a pair of scissors.

1. With your scissors, cut off a side-branch from the Busy Lizzie plant, cutting cleanly close to where the side-branch joins the main stem (Fig 5.8(a)). If possible choose a branch with no flowers on it. This is your cutting.

2. Fill the test-tube with water almost to the top, and stick your cutting into it as shown in Fig 5.8(b). Put it in a warm, well lit place, and leave it for a week or two. Top up the test-tube with water when necessary.

3. Observe your cutting every day. Watch for roots growing out of it. How long does it take for the first roots to appear? What part of the cutting do the roots grow from?

4. When several roots have grown out, remove the cutting from the test-tube and plant it in a pot of moist compost: make a hole in the compost with your finger, and carefully stick the cutting into it as shown in Fig 5.9. Press the soil round the cutting, taking care not to damage the roots. Leave it in a warm, well lit place and water the soil from time to time. If the cutting roots itself successfully, you will have obtained a new Busy Lizzie plant.

Questions for class discussion

1. Taking cuttings is an artificial way of reproducing plants, used by gardeners. In what circumstances might this kind of reproduction occur naturally?

2. Why is it best to take cuttings from branches which do not have any flowers on them?

3. Some gardeners take cuttings by sticking the branch straight into the soil; others put the branch in water until new roots start growing out and *then* they stick it in the soil. What are the advantages and disadvantages of each method?

Advantages and disadvantages of asexual reproduction

The most obvious advantage of asexual reproduction is that there is no need for the organism to have a partner. When asexual reproduction occurs by the formation of spores it occurs rapidly and enables the species to spread quickly over a wide area. Tubers and bulbs enable the plant to survive the winter.

The main disadvantage is that the offspring are always exactly like the parent. It is therefore impossible to produce new *kinds* of organisms by this method of reproduction. In other words asexual reproduction does not give rise to variety. Variety can only be produced by sexual reproduction, as we shall see in the next chapter.

Homework assignments
1. Four methods of asexual reproduction are listed below. In each case, write down the name of *one* organism which uses the method.
Binary fission, budding, spore-formation, tuber-formation.

2. Of the four methods of asexual reproduction listed in question 1, which ones:
a. enable the organism to survive the winter;
b. result in two or more of the offspring remaining attached to one another for a while;
c. produce two offspring every time reproduction occurs;
d. involve the formation of an underground structure containing food;
e. enable the species to spread quickly over a wide area?

3. The fern is a plant which has two methods of asexual reproduction: like the mushroom it can form numerous spores, and it also possesses an underground *rhizome* from which a new plant grows up each spring.

Suggest two advantages which the spores have over the rhizome as a means of reproduction.

Summary

1. Asexual reproduction is reproduction that does not involve two individuals.

2. Four of the main methods of asexual reproduction found amongst organisms are:
a. *binary fission* (e.g. *Amoeba*);
b. *spores* (e.g. mushroom);
c. *tubers* (e.g. potato);
d. *bulbs* (e.g. onion, daffodil);
e. *cuttings* (e.g. Busy Lizzie).
Taking *cuttings* is an artificial method of producing new plants used by gardeners.

3. The main advantages of asexual reproduction are that:
a. it does not require two individuals;
b. it may be rapid and result in wide dispersal;
c. it often enables the species to survive adverse conditions.

4. The main disadvantage of asexual reproduction is that it does not give rise to variety.

Chapter 6

Sexual reproduction

In Chapter 5 you saw how new individuals can be produced by a single parent, a process called *asexual reproduction*. But you probably know that in most organisms *two* parents are needed before young can be produced, and they have to be male and female. This method is called *sexual reproduction*. Each parent produces special sex cells known as *gametes* (pronounced 'gam-eet'). In animals, the gametes produced by the female are called *eggs* and those produced by the male are called *sperms*. In flowers, the male gametes are transported in the pollen grains. Offspring only develop if a male and a female gamete meet.

What happens when animals mate?

An important part of sexual reproduction in many animals is the process of *mating*. If the eggs and the sperms are released when the parents are very close together, there is a much better chance that a sperm will meet an egg. So, in many kinds of animal, the male climbs on to the female to mate with her. The photographs in Fig 6.1 show animals mating. In each photograph a male and female can be seen.

Questions for class discussion
1. Use the keys on pages 35 and 36 to name the group to which each pair of animals in Fig 6.1 belongs.

2. Describe the positions of the two animals in each photograph. What does this suggest about their sexes?

Fig 6.1 Animals mating

3. Look carefully for the differences in the appearance of the two animals in each photograph. Why do you think male and female animals often look different?

Experiment 6.1 Watching animals mating
If it is spring and you have a pond near your school, you may be lucky enough to see common frogs or toads mating.

Perhaps instead you may have locusts or African clawed toads in your laboratory. Under the right conditions they will mate at any time of the year.

1. Watch the mating animals very carefully without disturbing them. Write a clear description of mating, including details of the positions of the two animals. It may help if you make a sketch of them.

2. Amphibian eggs are called spawn. Place a small number in a preservative such as formalin or alcohol; you will be able to examine them in more detail later. Take eggs at various stages up to 24 hours after laying.

Questions for class discussion

1. Describe at least three differences in appearance between the male and the female.

2. What movements, sounds or responses did you observe before and during mating? Can you explain why they occurred?

3. What passed between the two animals while they were mating? Could you see any sign of this transfer taking place?

Eggs and sperms

When you watched animals mating you were probably unable to see anything passing between them. This is because their bodies were very close together to make sure that the eggs and sperms met. Eggs are rather large cells because they often contain a store of food called yolk. Sperms are very small and have a long tail so that they can swim to the egg. Figure 6.2 compares the eggs and the sperms of locust, hen and human.

Questions for class discussion

1. Use the scales on the drawings in Fig 6.2 to calculate the diameters of the eggs and the lengths of the sperms

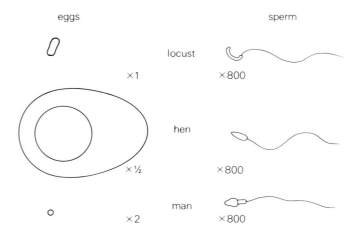

eggs

sperm

locust

×1   ×800

hen

×½   ×800

man

×2   ×800

Fig 6.2 Eggs and sperms of locust, hen and man

in micrometres (μm). Make a table in your notebook giving the sizes, and calculate how many times larger the egg is than the sperm.

2. Suggest a reason why sperms are so much smaller than eggs.

3. Apart from size, what other differences are there in appearance between eggs and sperms? Can you suggest how these differences relate to the functions of the egg and the sperm?

4. Name two structures possessed by eggs and sperms which are also found in other animal cells such as cheek cells (see page 50).

Fertilisation

When the sperm and the egg meet, a process called *fertilisation* takes place. The nucleus of the sperm combines with the nucleus of the egg. The fertilised egg is called a *zygote* (pronounced zy-goat). It forms the first cell of the *offspring*. A similar process takes place in flowers to produce the zygote inside a seed.

It is much more difficult to see fertilisation taking place than it is to watch animals mating, because the sperms can

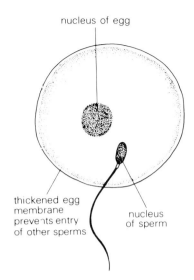

nucleus of egg

thickened egg membrane prevents entry of other sperms

nucleus of sperm

Fig 6.3 Fertilisation of a rat's egg

only just be seen under the highest power of a microscope. However, the drawing in Fig 6.3 shows what happens.

Eggs and sperms are not produced by the male and female in equal numbers. Millions of sperms are released for each egg even though only one will be needed to fertilise it. This increases the chances of successful fertilisation.

Where does fertilisation take place?

At certain times of the year, fish such as herring and cod are full of 'roe'. You may be able to obtain such a fish, or a small quantity of the roe it contains. The female roe is hard and lumpy and contains the eggs, while the male roe consists of a milky liquid which under a very powerful microscope can be seen to contain sperms.

The male and female fish gather in enormous shoals to breed, and they release their eggs and sperms into the sea. The sperms swim using their long tails until they find an egg to fertilise. Since the egg is fertilised outside the mother's body, we call this *external fertilisation*.

On land it would be difficult for fertilisation to take place outside the body of the mother since the sperms would be unable to move over dry ground or through the air. Reptiles, birds, mammals and some invertebrates such as insects have solved this problem by releasing their sperms directly into the body of the mother. The sperms swim to the egg and the result is *internal fertilisation*.

Questions for class discussion

1. List the animals shown in Fig 6.1. For each animal say whether fertilisation takes place inside or outside the mother's body.

2. An unfertilised frog's egg will start to develop into a tadpole if it is pricked with a fine needle. However, growth soon stops and the eggs never hatch. In *normal* fertilisation, what is needed to

a. start development?

b. continue development to the point of hatching?

Homework assignments

1. An egg is a special kind of cell. List the similarities and differences between a hen's egg and a human cheek cell (see page 50).

2. How many times larger is a hen's egg (Fig 6.2) than an onion cell 0.1 mm across? Can you suggest any reasons why onion cells cannot grow to the size of hens' eggs?

3. Explain briefly, in words which a non-biologist would understand, what you mean by *gamete*, *fertilisation* and *zygote*.

4. Name an animal in which fertilisation occurs inside the body, and another which has external fertilisation. Give one advantage of each method.

Similarities and differences explained

Asexual reproduction produces identical offspring, each one looking exactly like the others, and just like the parent. However, offspring produced by sexual reproduction are different. The advantage of this is that it provides an opportunity for new features to appear which may lead to the species being more successful in the future.

Why do brothers and sisters in a family differ from each other if the eggs and sperms came from the same mother and father? It is often said that a child 'takes after his mother' or that a particular characteristic has 'come from his father'. You can probably think of examples in your own family or your friends. In the family in Fig 6.4, which features in the children have been inherited from each parent?

From Chapter 3 you will remember that each cell is controlled by a nucleus. The nucleus contains the cell's instructions. When fertilisation takes place, a set of instructions from the father's sperm combines with a set of instructions from the mother's egg. If we think of the characteristics of the parents as resembling two separate packs of playing cards, the formation of eggs and sperms can be compared

Fig 6.4 A mixed-race family

with shuffling and dealing the cards. When fertilisation occurs the packs are put together again, all mixed up (Fig 6.5).

Questions for class discussion

1. Stick insects are unusual because they produce eggs which can develop directly into young without fertilisation. Discuss the advantages and disadvantages of this method of reproduction.

2. Three of the most difficult weeds to remove from a garden all reproduce asexually from underground parts of the plant. They are dandelion, mare's tail and couch grass. However, once they have been eliminated from a garden completely they can be kept out more easily than sexually reproducing species such as chickweed. Why do you think this is?

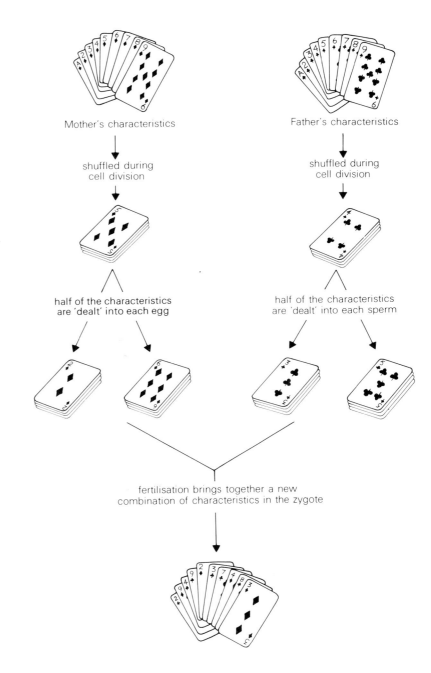

Mother's characteristics

Father's characteristics

shuffled during
cell division

shuffled during
cell division

half of the characteristics
are 'dealt' into each egg

half of the characteristics
are 'dealt' into each sperm

fertilisation brings together a new
combination of characteristics in the zygote

Fig 6.5 How characteristics are
inherited

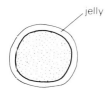

newly-fertilised egg

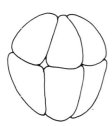

embryo: 8 cells

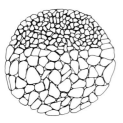

embryo: many cells

Fig 6.6 Early stages of growth in *Xenopus* eggs

Growth

What changes take place as a newly fertilised egg (zygote) turns into a fully grown organism? The most obvious change is the increase in size (*growth*) which is necessary.

An elephant's zygote is about 0.1 millimetres across; the adult elephant is about 5000 millimetres long. How many times longer is the adult than the zygote? Assume that the elephant is also 5000 millimetres tall and broad i.e. a cube. Calculate the volume of the elephant and compare it with that of the egg. Now you have some idea of the enormous growth which takes place, and the number of new cells which are required as the zygote becomes an adult elephant.

As the zygote divides and grows it is called an *embryo*. The early stages can be seen quite easily in frog or toad eggs.

Experiment 6.2   Watching the zygote divide

The early stages of division are visible in the eggs of frogs or toads. A microscope or lens can be used, but a binocular microscope will give the clearest view.

1. Place a few eggs into a watch glass or cavity slide, using a wide-mouthed pipette. Suck them up gently or they will be damaged. There is no need to add a coverslip.

2. Examine them under the lens or low power of the microscope.

3. Try to find and draw newly laid eggs, 2-cell, 4-cell and 8-cell embryos and one or two later stages (Fig 6.6).

Questions for class discussion

1. The upper surface of a frog's egg is dark, but the lower surface is light. What advantages are there in this pattern?

2. Most frog and toad spawn contains some eggs which do not divide and grow. How can you recognise these eggs and what proportion of the total number do they represent? Suggest reasons for the failure of these eggs to grow normally.

Development

If you watch the growth of amphibian eggs for 24–48 hours, you will begin to see new changes which cannot be explained by growth and increase in the number of cells alone. The round shape of the egg begins to change into the sausage shape of the tadpole. If you look carefully you will see new features start to appear on the surface. Inside the body, the heart and blood are forming. These changes, which we refer to as *development*, begin soon after the egg is fertilised.

Some organisms develop steadily towards adult shape, size and structure. For example, a baby has most of the features of an adult human being long before it is born. However, young frogs and toads do not resemble adults. A tadpole looks very different from the frog into which it will develop. We call the tadpole a *larva*.

Caterpillars and maggots are examples of the larval stage of insects, but they do not turn directly into adults. Instead, the change from larva to adult takes place inside a protective case. This stage is called the *pupa*.

If you keep your tadpoles for 2–3 months, you will probably see the tremendous change which takes place as the tadpole turns into a frog or toad. A sudden change during development is called *metamorphosis*. The larval structures are broken down and replaced by the adult structures. Different types of development in man, frog and butterfly are shown in the form of a *life-cycle* in Fig 6.7.

The larval stage of the life-cycle is a time of feeding and growth. However, larvae often have little protection from their enemies. So why do many animals include a larval stage in their life-cycles?

One of the reasons may be to reduce competition between parents and offspring for the same kind of food. Tadpoles feed mainly on plants in the water while their parents eat small animals on land. Caterpillars and butterflies also feed in completely different ways. As you will see in the next chapter, baby locusts have to compete with their parents for food because their life-cycle does not include a complete metamorphosis.

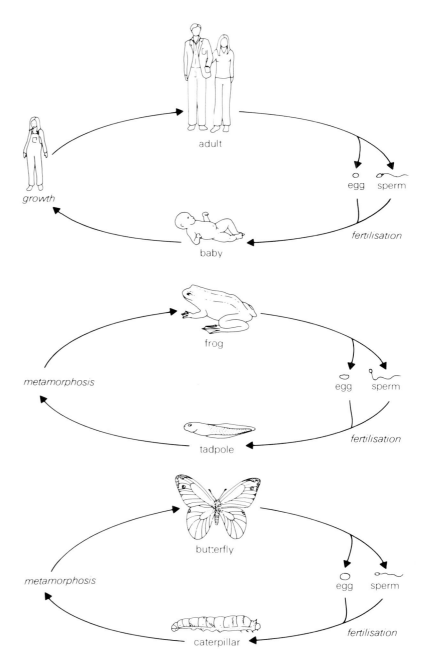

Fig 6.7 Life-cycles of man, frog and butterfly

Questions for class discussion

1. Food is needed for growth. Where does the fertilised frog's egg obtain the food which it needs if it is to divide and grow?

2. If you take a dividing frog's egg and carefully separate the first two cells which are produced, they will both develop into normal tadpoles. Explain as clearly as you can the reasons why

a. the two tadpoles will have identical characteristics

b. the tadpoles develop more slowly and are smaller than other fertilised eggs which were not separated into two cells.

Homework assignments

1. Compare sexual and asexual reproduction giving suitable examples. Explain briefly the advantages and disadvantages of each method.

2. What is meant by the term metamorphosis? Name *two* animals belonging to different groups which have a larval stage in their life-cycle. What advantage does the larval stage give them?

3. Answer the following questions from your knowledge of frogs, toads or locusts.

a. How can males and females be distinguished?

b. What behaviour takes place *before* mating?

c. Describe briefly how and where they mate.

d. Is fertilisation internal or external?

e. Roughly how many eggs are laid?

f. What are the functions of the material which surrounds the egg?

g. Describe and explain the appearance of the newly laid egg.

h. How many days do the eggs take to hatch?

i. List the changes which take place in structure and behaviour as the newly hatched offspring develops into an adult.

4. Write out each of the following sentences. Choose the *best* word from the list to complete each sentence.

a. An egg can only develop into an adult after _____ .
b. _____ is an important part of sexual reproduction in all land animals.
c. The change from larva to adult is called _____
words: metamorphosis
mating
fertilisation

5. Sperms and eggs are special sex cells. What general name do we give to such cells? Describe three ways in which sperm cells and egg cells are different and one way in which they are similar.

Background reading

The curious sea-horse
Sea-horses are primarily inhabitants of the warmer seas. In Europe they occur as far north as in the English Channel, but they begin to be rare in these latitudes, whereas they are very common in the Mediterranean Sea. The most abundant varieties do not grow larger than about five centimetres but the tropical forms sometimes attain much greater size. In the Australian Seas and also near Japan large specimens occur. The largest are half a metre in length.

Every aspect of these strange creatures – their appearance, movement, feeding and breathing – is unusual, but none more so than their reproduction.

The adult female carries about two hundred eggs. When the time has come for mating, the male and female approach each other and begin to make movements which may be compared to a dance.

The male is equipped with a ventral pocket or pouch with a small slot-like opening. The female inserts her cloaca into this slot and projects her eggs into the pouch; while passing the slot they become fertilised.

When the pouch of the male sea-horse is not occupied, it is small and wrinkled. But from the moment the eggs are deposited a considerable change takes place. The tissue begins to swell and grow, and the blood vessels enlarge and multiply.

Soon after the eggs have entered the pouch, each one produces a localised excitation and is engulfed in a separate compartment. The shell, or rather the skin, of each egg splits open inside the pocket, but the embryo is not yet expelled. It remains in the pouch until its yolk is used up almost entirely.

Forty to fifty days after fertilisation, the young ones are expelled. The male appears to suffer considerable difficulty in the act of expulsion. He can be seen writhing on the sand, rubbing his body, and struggling energetically. Finally, with considerable force, the male ejects his burden of young sea-horses.

(From *Strangest Creatures on Earth* edited by Edward H. Meyer, published by Harrap.)

Questions

1. Describe *two* ways in which fertilisation in the sea-horse differs from fertilisation in most other animals.

2. For thousands of years it was assumed that the *female* sea-horse 'incubated' the eggs. It was not until cine cameras were available that it was confirmed that it was the job of the male. Explain exactly what evidence you would need to prove that it was the male that had the pouch.

3. What are the *two* sources of food for the developing embryo?

4. Why do you think the male has 'considerable difficulty' when the young sea-horses leave his pouch?

Summary

1. Most animals possess male and female sexes, and reproduce by *sexual reproduction*.

2. Sex cells are called *gametes*. *Sperms* are produced by the male and *eggs* are produced by the female. An egg and a sperm combine in a process called *fertilisation* to form a *zygote*.

3. The zygote *grows* by cell division to form an *embryo* which *develops* into an adult.

4. Some embryos develop steadily into an adult; others first form a *larva* which undergoes a sudden change or *metamorphosis* into an adult.

5. The main advantage of sexual reproduction over asexual reproduction is that the offspring differ from each other and differ from their parents.

Chapter 7 # Animal life-cycles

In this chapter we shall study the life-cycles of four groups of animals: insects, amphibians, birds and mammals. If your school has facilities for keeping some of these animals, or if you can keep them at home, you will be able to study them yourself. If your animals are different from the ones described in this book, compare them with the animals described here. Try to suggest reasons for any differences you find between your observations and measurements and those given here.

Insect life-cycles

The phylum Arthropoda, to which insects belong, includes almost nine-tenths of all the animal species known. 'Arthropod' means 'jointed foot', and this is one of their main characteristics. They also have a hard outer covering and a soft 'inside'. Insects are easily distinguished from the other groups of Arthropods because they have three pairs of jointed legs. Spiders have four pairs, crustaceans (e.g. woodlice, crabs and lobsters) have between five and seven pairs and centipedes and millipedes have many more.

For every species of mammal there are over 200 insect species and they are found in almost every type of habitat. The dried larva of a West African midge can survive temperatures from 100°C to −200°C for several hours! Without natural enemies and given unlimited food, one winter egg of a greenfly could, by next autumn, produce enough offspring to equal the mass of the Earth. No wonder they are so widespread! Many insects are harmful to man. Locusts destroy his crops and mosquitoes carry serious diseases such as malaria. A few species have been used by man for many years for example, silk worms and honey bees.

Experiment 7.1 Studying the locust

Locusts are among the most serious pests in the world. A swarm of locusts can destroy millions of hectares of crops in a few hours. Two species are often kept in captivity – *Locusta migratoria* (the African Migratory Locust) and *Schistocerca gregaria* (the Desert Locust). They are easily distinguished by their size and colour (Table 7.1, Fig 7.1).

Table 7.1

Fig 7.1 (a) *Locusta migratoria*
(b) *Schistocerca gregaria*

Species	Where it occurs	Usual colour	Usual mass
Locusta migratoria	Tropical Africa	Brown	1.0g (Male) to 1.5g (Female)
Schistocerca gregaria	Central and North Africa and Asia	Male: yellow Female: pink	1.3g (Male) to 2.0g (Female)

1. Transfer a living locust from the cage to a small container or dish. To pick up a locust hold it between your thumb and forefinger across the top of the thorax. This prevents the wings from opening and allows you to look at the whole insect closely. Locusts become quieter and easier to handle when they are cool, so turn off the lights in the cage a short time before you wish to take them out.

2. Use Fig 7.2 to discover whether the locust you are examining is male or female. Compare the structure of both sexes.

3. Use Fig 7.2 to identify the other main parts of the body and answer these questions.
a. Describe the 'feel' of the locust. Is it hard or soft, wet or dry, rough or smooth?
b. Compare the front and hind wings, unfolding them carefully. Locusts will sometimes try to fly if you hold them with their feet free and blow gently on their heads.
c. Examine the three pairs of legs. How do the hind legs differ from the others? Can you suggest how the locust uses the features you describe?

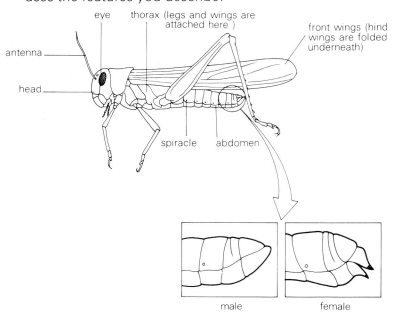

Fig 7.2 Structure of a locust

d. Find the spiracles through which air enters the body. Can you see them opening and closing? Insects have air tubes running throughout their bodies carrying air to the cells. If you place the locust in a boiling tube and breathe into the tube, what happens to the breathing movements?

e. How does the locust see? How does the locust detect movement of the air?

4. Return the locust to its cage and watch it feeding. Make your own notes, with sketches, explaining how it tests, holds and bites its food. If you have any dead locusts available, use a mounted needle and hand lens to examine the mouth parts, including the large, horny jaws.

Questions for class discussion

1. Discuss the ways in which the structure of a locust is adapted to its way of life.

2. Describe the main features of the cage in which your locusts are kept. Explain what each part of the cage is for.

Experiment 7.2   Breeding locusts

Locusts breed readily in the laboratory if

a. the temperature is kept at 28°C to 30°C

b. they are well fed on a variety of green foods

c. there are sufficient numbers of each sex in the cage

d. they are provided with fresh damp sand at regular intervals.

Remove the lid from a small sandwich box or yoghurt carton not less than 7 cm deep. Half fill the box with dry sand and then add 1 part water for every 10 parts of sand before placing it in the cage. If a little white sand is placed on top, it is possible to see when the eggs have been laid in the box (Fig 7.3).

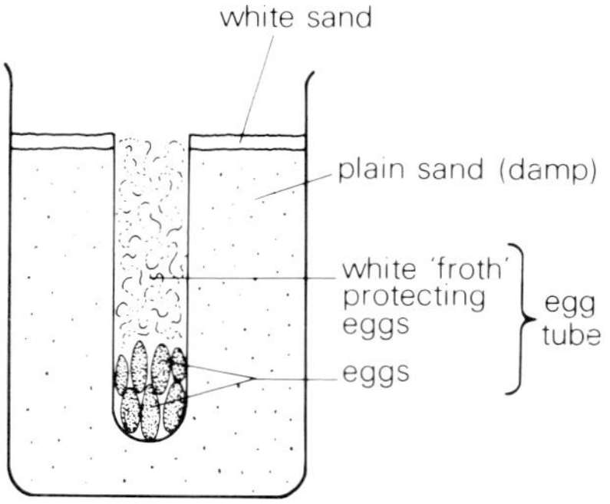

Fig 7.3 Locust eggs laid in sand

Experiment 7.3   Measuring the growth of locusts

The eggs should hatch after about eleven days. Locusts

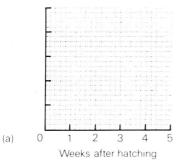

(a)

Weeks after hatching

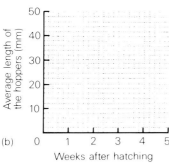

(b)

Weeks after hatching

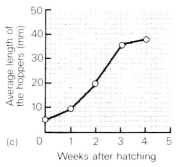

(c)

Weeks after hatching

Fig 7.4 Drawing graphs of growth

have no larval stage in their life-cycle. The newly hatched locusts (called *hoppers*) look like small adults with no wings. The hopper spends a few days feeding, and then its hard outer covering splits and it moults. The new covering is soft at first, allowing the locust to expand and grow before becoming hard again. The locust has five moults before the adult emerges. This kind of development is sometimes called *gradual* or *incomplete metamorphosis*.

1. The day before you expect the hoppers to emerge, remove the lid from the box containing the egg tubes and place the box in an empty locust cage in which there is soft grass and a small dish of bran to eat. Twigs should be provided for the hoppers to cling to when they moult.

2. Prepare a table in your notebook with these headings.

Table 7.2

Date	Temperature (°C)	Number of locusts measured	Average length (mm)	Other observations

3. Check your locusts as often as you can. (In the first week the locust hoppers will be too small to handle so measurements should be estimated.) At each visit you should record

a. the temperature to the nearest 1.0°C;

b. the number of locusts you measure (not less than 5, taken at random);

c. their average length (see Experiment 2.1 if you cannot remember how to work out the average). Place them in a dish on top of a piece of millimetre-squared graph paper;

d. changes in their appearance or their behaviour, number of deaths, and the amount and type of food which they eat.

4. Interesting results can be obtained if the temperature is altered by 2°C or 3°C above or below the normal value, or if different types of food are provided.

5. After the first week, clean out the old grass and supply fresh regularly. Check the old grass for hoppers before you throw it away! When the hoppers are small it may be best to empty the cage inside a large box to reduce the risk of losing any. Watch out for the moulted skins of the locusts and collect any you find. You could stick them carefully into your notebook.

6. Draw a graph of your results as follows. Plot the dates, from hatching to your final reading, along the horizontal axis. If possible, show *all* the days and not just the ones when measurements were recorded. (Fig 7.4 (a)).

7. Write 'Average length of the hoppers (mm)' along the vertical axis. (Fig 7.4(b)).

8. Mark each of your results in your table and join the points with a neat line (Fig 7.4(c)). Finally, give the graph a title including the *average* temperature; for example, 'Graph to show the growth of *Locusta migratoria* at 26°C'.

Questions for class discussion

1. A newly-hatched locust is pale in colour at first, but soon becomes darker. Can you suggest what is happening to the skin at this time?

2. Describe what you observe happening when a locust moults. Why is moulting necessary?

3. When do wings first appear on your developing locusts? How do the hoppers manage without wings until then?

Homework assignments

1. The stages in the development of a locust are called *instars*. If the instars of a single locust are measured at frequent intervals from hatching, the graph shown in Fig 7.5 might be obtained.

a. What causes the sudden increases in length?

b. Why did *your* graph not show these sudden increases?

2. Locusts sometimes form enormous swarms which can be hundreds of kilometres wide. As the swarm moves, the locusts destroy all the crops in their path. Use your

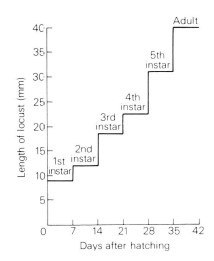

Fig 7.5 Growth and development of a locust

school library or books at home to find out as much as you can about the harm caused by locusts and the methods used to destroy them.

3. The life-cycle of a butterfly is shown in Fig 6.7. Can you draw a similar diagram to show the life-cycle of a locust? (Remember: locusts only have an incomplete metamorphosis in their development.) How long does the locust take to complete its life-cycle?

4. What changes take place during the pupal stage of the life-cycle of a butterfly? Where do similiar changes take place in the life-cycle of a locust?

Amphibian life-cycles

Amphibians, such as frogs and toads, have a soft body with a bony skeleton in the middle – the backbone. Animals with backbones are called vertebrates. In spite of their name, some amphibians spend all their lives in water, while others remain mostly on land. However, even the amphibians which are most at home on dry land have to return to the water to breed.

The most common British frog is *Rana temporaria* and the most common toad is *Bufo bufo* (Fig 7.6). They both have a damp skin, which means that they are restricted to wet places, but the common frog's skin is smooth while the common toad's skin is rough. In the autumn they *hibernate* in the mud or under stones. They stop moving and their heart beat slows down to save energy. In spring they emerge to breed in ponds, rivers or canals.

Experiment 7.4 Studying frogs and toads
Unfortunately, the common frog is becoming less common in many parts of Britain as the land is more intensively cultivated and developed. Only small quantities of spawn should be brought into the laboratory, and the tadpoles or small frogs should be returned to the pond afterwards. A related species, the African Clawed Toad *Xenopus laevis* is much easier to keep, requires no heat except when breeding,

Fig 7.6 (a) The common frog *Rana temporaria*   (b) The toad *Bufo bufo*

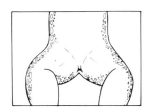

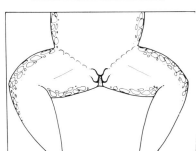

Fig 7.7 Dorsal surface of abdomens of male (above) and female *Xenopus* toads

and can go for long periods without feeding or cleaning. Full instructions for keeping Xenopus will be found in a number of reference books. You will need a mature pair of *Xenopus* toads to carry out the following observations.

1. Observe the male and female toads carefully and list the differences between them. A fully grown adult female is almost twice the size of a male and much broader in the body. Figure 7.7 also shows other features which are useful in distinguishing the sexes.

2. Remove the cover from the tank and carefully catch one of the toads in an aquarium net. What does the skin of the toad feel like? *Xenopus* toads do not normally leave the water and will die if they are removed for more than a few hours. Try to explain why this happens.

3. Take two containers and fill them with water. Place a toad in each dish and cover with a sheet of glass. Enclose one of the dishes in black cloth and leave the other standing on a piece of white paper illuminated by a bench lamp. Compare the appearance of the two toads after about twenty minutes and describe the ability of *Xenopus* to change colour. This is called *camouflage*.

4. Hold a small piece of meat in front of one of the toads to investigate its response to food. Bring it gradually closer to the toad's nose until it catches it. Describe the way in which the toad finds and captures its food.

5. Tap gently on the table or side of the tank. Does the toad respond to vibration and if so how? African Clawed Toads live in muddy pools and ponds in the southern half of Africa. How do the senses which you have investigated help them in their natural habitat?

6. Describe *three* ways in which the toad is adapted to swimming.

7. Look carefully at the toads when they are still. Can you see them breathing? Gently touch one of the toads with the aquarium net so that it swims round the tank several times. Look again for signs of breathing. Describe your observations and try to explain them.

Experiment 7.5   Measuring the growth of Xenopus tadpoles

British frogs and toads mate and lay eggs from February onwards in Southern Britain and from May onwards in the North. For a few days each spring, ponds may be alive with them, and their mating calls can be heard a long way away. The African Clawed Toad, *Xenopus laevis*, will lay eggs at any time of the year. You may have studied the early stages of development of the fertilised eggs in the previous chapter (Experiment 6.2). The *Xenopus* tadpoles are easy to rear and develop rapidly if (a) they are kept at about 23°C, (b) they are fed sparingly, (c) their water is changed regularly.

1. Take about 20 newly hatched *Xenopus* tadpoles and place them in fresh pond water, or tap water which has stood for 24 hours. Several containers can stand in one tank containing water maintained at 23°C by a thermostatically controlled heater.

2. Prepare a table for your results (see Experiment 7.3). The tadpoles can be measured by capturing them in a fine aquarium net, and placing them in a drop of water in a

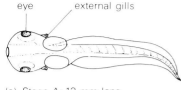

eye   external gills

(a) Stage A: 13 mm long

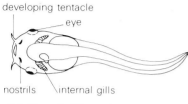

developing tentacle

eye

nostrils   internal gills

(b) Stage B: 28 mm long

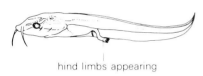

hind limbs appearing

(c) Stage C: 65 mm long

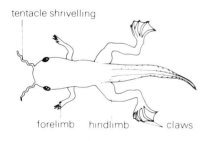

tentacle shrivelling

forelimb   hindlimb   claws

(d) Stage D: metamorphosis

Fig 7.8 Stages in the development of *Xenopus* toads

glass dish on top of millimetre graph paper. Use a lens to examine them in detail, and make a drawing once every week. The main stages of development are shown in Fig 7.8.

3. Check and record the temperature each time you measure the tadpoles. Their natural food when first hatched is microscopic plants but they will grow rapidly if they are fed on a teaspoonful of baby food enclosed in a bag of fine muslin and hung in the water for about 10 minutes each day. Top up the water to the original level, or change the water when the tadpoles have finished feeding.

4. After a month or so the tadpoles begin to change their appearance and behaviour as *metamorphosis* takes place and the *larva* (tadpole) becomes an *adult*. The first sign is the appearance of hind legs. Later the front legs appear. Gradually you should give them more animal food such as tiny worms or finely chopped meat. You will also see them come to the surface for air. Continue your regular measurements and watch out for the changes listed above, noting when they first appear.

5. Carry out the same instructions that were given for locusts in Experiment 7.2 and draw a graph to show the growth of *Xenopus* tadpoles.

Questions for class discussion
1. When the tadpoles first hatch they have simple external gills for breathing, just behind the eyes. Later on they take water into their bodies and use more complex internal gills. Can you suggest why they need more air as they grow larger?

2. Both tadpoles and caterpillars are larval stages of animals that have metamorphosis in their life-cycles. Explain what is meant by metamorphosis and describe the part played by the larva in the life-cycle.

Homework assignments
1. Draw a simple life-cycle diagram for a frog or toad, based

on Fig 6.7. Illustrate your cycle with a sketch of each of the main stages. How long does this species take to complete its life-cycle?

2. List as many differences as you can between the structure and behaviour of a tadpole and the adult frog or toad which you have studied.

3. There is usually a considerable variation in the rate of growth and development of tadpoles, even when they have come from a single batch of spawn kept under the same conditions. How many of your tadpoles are at each of the stages of development shown in Fig 7.8? What might cause this variation, and why is it an advantage to the toads not to have all the tadpoles growing at the same rate?

4. It has been suggested that larger tadpoles release a very small quantity of a chemical substance into the water which slows down the growth and development of smaller tadpoles. How would you test the idea experimentally? (*Note*: the quantity of chemical is too small for identification or extraction.)

The life-cycle of a bird

Birds are very successful land animals. They can survive in most types of habitat on land, and a few species have even returned to live in the water. Their success on land is due to, amongst other things, their control of body temperature (about 40°C or 104°F), restriction of water loss, ability to fly, and their life-cycle, which includes internal fertilisation.

Experiment 7.6 What is in an egg?
The main parts of a newly laid hen's egg are shown in Fig 7.9.

Each part has a special job to do. If an egg is laid on land, a hard outer *shell* is important to protect it and reduce

Fig 7.9 Vertical section of a hen's egg

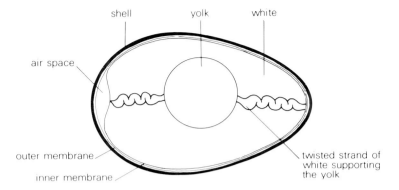

shell   yolk   white

air space

outer membrane

inner membrane

twisted strand of white supporting the yolk

Fig 7.10

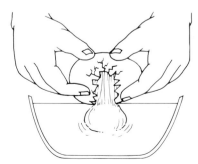

(a) Cracking open an egg

(b) Emptying out the contents

evaporation of water. The two *membranes* inside the shell also help to control water loss, but allow air to pass through. The *white* of the egg protects the embryo and provides it with water, and the *yolk* feeds it.

1. Half fill a bowl or basin with water.

2. Holding the egg at the two ends, tap the centre sharply on the side of the bowl (see Fig 7.10(a)).

3. Hold the egg so that it touches the surface of the liquid in the bowl, and break the egg in half. Allow the contents to empty out gently (see Fig 7.10(b)).

4. Look at the membranes in the blunt end of the egg shell. The air space should be clearly visible now. Use a pair of forceps to remove a small piece of the inner membrane. Break the shell to obtain a piece of the outer membrane. Are they different? Which is thicker?

5. Look in the bowl at the clear egg white. Can you see the twisted white strands which attach the yolk to the shell membrane? They allow the yolk (and embryo) to turn over.

6. Look carefully at the yolk, turning it over if necessary until you can see a small round patch on the surface. This consists of the cells from which the embryo would be formed if the egg had been fertilised. You will see what happens to them in Experiment 7.7.

Questions for class discussion

1. Which is the larger part of the egg, the yolk or the white? How could you test your answer experimentally? What jobs do the yolk and the white perform?

2. Compare a bird's egg with an amphibian's egg. List and discuss the similarities and differences.

3. If you remove the shell from a hard-boiled egg, a slight hollow can often be seen at the blunt end of the white. Can you explain what it is, and what job it performs in the developing egg? Why is it that piercing the blunt end of an egg with a pin will stop it cracking when it is boiled?

Experiment 7.7   Incubating fertilised hens' eggs

The eggs you examined in Experiment 7.6 did not contain an embryo because they had not been fertilised. Fertile eggs can be bought from hatcheries, and will develop and hatch if placed in an *incubator*. The main features of an incubator are shown in Fig 7.11.

1. Write the date on the side of the fertile eggs when you start to incubate them. Place them in the incubator with the date uppermost. Check the temperature. It should be approximately the temperature of a hen's body, 40°C or 104°F. Top up the water trough to ensure that the air in the incubator is not too dry.

2. Check the temperature and water as frequently as you can. Turn the eggs over at each visit to make sure the temperature is even throughout the egg.

3. Open an egg which has been incubated for 3–5 days by cracking it into warm water (see Experiment 7.6 for the method).

4. Examine and draw the embryo, which is attached to the upper surface of the yolk. The use of a lens and a bench lamp shining on the egg from the side will make the details clearer.

Until the egg has been incubated for 24 hours you will

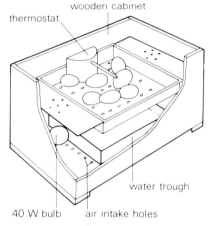

thermostat

wooden cabinet

water trough

40 W bulb   air intake holes

Fig 7.11 An incubator for hens' eggs

see little sign of the embryo. Later, the appearance changes quickly if the egg has been fertilised.

5. The time taken for the eggs to hatch is called the *incubation period*. You may not have the time or the facilities to incubate your eggs until they hatch. The pictures in Fig 7.12 show what happens. By this time, most of the yolk will have been used up by the embryo. A hook develops on the top of the beak called an *egg-tooth*. Using this hook, the chick breaks through into the air space and then cracks the shell by tapping it. It then pulls itself out of the broken shell and soon dries out. The membranes and waste products are left behind in the shell. The baby bird will need to be kept warm, fed and protected for several weeks after hatching.

Questions for class discussion or homework

1. Discuss the ways in which the following features help birds to live on land:
a. feathers;
b. a constant body temperature;
c. internal fertilisation;
d. shelled eggs;
e. incubation of eggs;
f. care of young.

2. List the conditions that are required by a developing hen's egg. Explain briefly why each condition is necessary and how an incubator provides it.

3. A widely used method of distinguishing between fertile and infertile eggs involves placing them in a bowl of water. Those that float are said to be fertile, those that sink, infertile. Using your knowledge of the structure of an egg, and the changes that take place as it develops, can you suggest a reason for the difference?

4. Most birds lay their eggs at intervals of a day or more, yet they usually hatch more or less together. Recent studies have suggested that, in some species, this may be due to communication between the unhatched chicks.

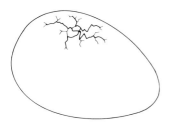
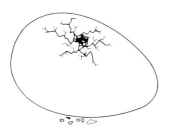

Fig 7.12 The hatching of a chicken

a. By what means might they communicate?

b. How would you test the idea?

5. Draw a simple life-cycle diagram for a bird, based on Figure 6.7. How long does the hen take to complete its life-cycle?

Experiment 7.8 Studying birds in the wild

Bird watching is one of the most popular branches of natural history. All that is required is a good eye for detail. You can make your interest in bird watching even more useful if you join a club, take part in local or nation-wide investigations or simply record your own observations. Many good books are available both for identification and further studies (see Appendix), but here are some suggestions from which a beginner might choose a short project.

1. A bird table or even bags of nuts hung up outside a window can provide a fascinating source of information, especially in winter. Set aside a few minutes at the same time of each day and record how many individuals, of which species and sex, visit the table and for how long. Which food do they take, how do they collect it and how do they respond to the presence of other birds? The visiting birds can be related to the weather conditions (remember to record them), amount and type of food provided, position of food (high, low etc.) time of day and so on.

2. Watching a nest can provide much useful information, but remember that some birds are very easily disturbed. If you can watch the nest from a window, or from a place where people pass regularly, then you are unlikely to risk disturbing the birds. Most birds will put up with being watched more readily when they are feeding young than when they are building a nest or incubating eggs. Again the secret is to watch regularly for a short period each day. Record the number of visits made by each parent, what food they bring and how long they stay. Relate this information to the number and age of young birds, weather conditions and time of day.

Fig 7.13 Song thrush feeding its young

3. A population study requires a little more knowledge of birds and, ideally, a pair of binoculars. Useful information can be obtained, even if only the most common species are recorded. The best method is to take a regular walk along the same track at the same time each day. Make a sketch map of the area or copy a large-scale map, and plot the places where any birds are seen or heard. This can be a particularly valuable and interesting study early in the breeding season if you mark down the exact positions of singing birds. Since most birds only sing in their nesting territories, such a study gives a good indication of the nesting population. You can then relate the size of the territories to the type of habitat.

The life-cycle of a mammal

Mammals, like birds, are 'warm-blooded' animals – that is they maintain a constant body temperature. However, instead of having feathers to reduce the heat loss, mammals are covered with hair. The life-cycle of a mammal differs from that of a bird because the fertilised eggs stay inside the mother's body until the embryo is well developed. This is called *internal development* and the time taken from fertilisation to birth is called the *gestation period*. After birth the baby mammals are fed by the mother on a special milk which is produced in her mammary glands, and they are protected by their parents until they are completely independent.

Most of you will be familiar with one or more species of mammal. You may have small mammals such as rats, mice, gerbils or hamsters at home or in school. Many homes will have a cat or a dog. If you live on, or visit, a farm you may be familiar with cows, sheep or horses. Choose any one mammal species, and find out the answers to the following questions.

1. Is there any simple way of distinguishing which is a male and which is a female?

2. What food does it require and how does it feed?

3. What is its natural food in the wild, and in which parts of the world did it originate?

4. In what ways is it suited to its natural habitat and way of life?

5. Which animals (if any) would prey on it? What defences does it have against attack?

6. How long does it live?

7. How many babies are produced at a time?

8. How long do the young take to develop into adults?

9. How does it keep warm in the night and cool in the daytime?

10. In its natural habitat does it live alone, in pairs or in a colony?

Experiment 7.9 Studying the behaviour of mammals

Any family of young mammals can be used, but the Mongolian gerbil (*Meriones unguiculatus*) is particularly easy to keep and breed. The babies of all small mammals should be left undisturbed until they stop feeding from their mother. When this stage is reached, the baby is said to be *weaned*. The temperature should be kept at 18°C–20°C; a bench lamp will supply the right amount of heat at night or in the winter. The gerbils will tear up paper for nesting material, and a thick layer of sawdust will help keep them clean.

1. Check the animals regularly; refill their food and water containers, but do not attempt to clean the cage until the babies are weaned.

2. Record very carefully what you can see the babies doing each time you visit them. Also note what the parents are doing.

Fig 7.14 Mongolian gerbils

3. After how many days from the birth of the babies
a. do the eyes open
b. do they start to leave the nest
c. do they begin to feed for themselves?

Questions for class discussion or homework

1. Make a list of the main changes in structure and behaviour which take place as a young mammal develops.
2. Draw a graph of the changes in mass, shown in Table 7.3, of a family of young mice. Use your graph to answer the questions which follow.

Table 7.3

Days after birth	0	5	10	15	20	25	30	35	40
Average mass (g)	1.5	3.0	5.0	7.5	10.0	15.0	19.0	21.0	22.5

a. How long did it take the baby mice to double their mass after the first measurements?
b. How long did it take them to double their mass in the period ending in the last measurement?
c. What does this tell you about the rate of growth of the young mice at different stages of development?

3. Draw a simple life-cycle diagram for a named mammal, based on Fig 6.7. How long does it take to complete its life-cycle?

Homework assignments

1. Table 7.4 shows the mass and length of chick embryos during incubation.

Table 7.4

Days	3	5	7	10	12	14	16	18	20
Mass (g)	0.02	0.13	0.57	2.26	5.07	9.74	15.98	21.83	30.21
Length (mm)	12	15	20	35	50	65	75	80	82

a. Plot a graph to show the increase in mass of the chick, putting days of incubation along the horizontal axis.
b. When did the embryo increase its mass most rapidly?

c. When did the length of the embryo increase most rapidly?
d. How would you measure   (i) the length   (ii) the mass of the chick embryos?

2. The time taken for a mammal to develop inside the body of its mother from fertilisation to birth is called the gestation period. Make a table to show the gestation periods for two mammals and the incubation periods for two birds.

3. Copy and complete Table 7.5, which compares the ways in which the developing embryos of the five groups of vertebrates are provided for.

Table 7.5

Animal group	Food	Water	Air	Warmth	Protection
Fish					
Amphibians					
Reptiles	Yolk	White	Shell	Environment	White and shell
Birds					
Mammals					

4. List the problems which animals have to overcome when they live on land rather than in the water. Describe briefly how each problem is overcome in
a. birds
b. mammals.

5.
a. Name one particular kind of animal, with a life history that you have studied, which does not change its form (metamorphose) during development.
b. How long does it take, from hatching or birth, for it to become adult?
c. Give a clear description of the adult form, including details of its size, shape and colouring. (Make a labelled drawing if this helps.)
d. How can you distinguish between the adult males and females? Give details.
e. If the female lays eggs, say how many and how often she lays. If not, explain briefly how offspring are produced.

f. How are the offspring fed and how are they protected?
g. Upon what kind(s) of food does the adult depend?

6.
a. Many kinds of animals in Britain reproduce only in spring and summer. Give an example. What are the benefits of this?
b. Some animals in Britain hibernate in winter. Give an example. What are the benefits of this?

7.
a. Name an animal you have studied with a life history that shows a change of form (metamorphosis).
b. State *three* ways in which the larva differs from the adult.
c. In what natural habitat would you be most likely to find   (i) the larva   (ii) the adult?
d. State one advantage and one disadvantage of a life-cycle which includes metamorphosis.

8.
a. The graph in Fig 7.15 shows a major difference between two animals. What is this difference?

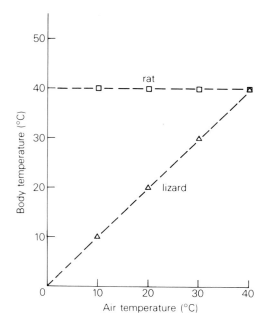

Fig 7.15 Comparing the body temperatures of two vertebrates

b. Which animal is more likely to remain active in both hot and cold surroundings? Give reasons for your answer.

9. A boy decided to test the effect of temperature on the growth of tadpoles of the African Clawed Toad by keeping some in each of two tanks, A at 20°C and B at 24°C. The boy found that the temperature was not the only factor affecting the growth of the tadpoles.
   a. List *five* other factors which might affect their growth.
   b. Why was it necessary to have two tanks of tadpoles at different temperatures?
   c. What method would you use for (i) measuring the tadpoles (ii) calculating the average growth of the tadpoles in each tank?
   d. Describe, in outline, what you would expect to happen in the two tanks.
   e. During the experiment several tadpoles died. Describe *two* ways in which their deaths might upset the experiment.

Summary

Table 7.6 summarises the main facts about the four animals and their life cycles which have been described in this chapter.

Table 7.6

Animal	Group	Warm or cold blooded	Type of egg	Laid in	Fertilisation	Developmemt of embryo	Metamorphosis	Care of young
Locust	Insect	Cold	Stiff shell	Land	Internal	External	Incomplete	None
Xenopus toad	Amphibian	Cold	No shell – jelly	Water	External	External	Yes	None
Hen	Bird	Warm	Large, yolky, shelled	Land	Internal	External	No	Yes
Gerbil	Mammal	Warm	Small, no yolk or shell	—	Internal	Internal	No	Yes

Chapter 8 **Plant nutrition**

You and I feed by taking in solid food substances. However, plants do not do this. Have you ever wondered how they feed? This chapter is all about how plants feed.

The leaves of green plants normally contain *starch*. We can prove this by testing a leaf with iodine. If starch is mixed with iodine, it goes a blue-black colour. However, to see this colour change in a leaf it is necessary to remove the green pigment first.

Experiment 8.1   Testing a leaf for starch

The test should be carried out on the leaf of a potted plant such as a geranium. The plant should have been kept in a well lit place for several days beforehand.

1. Half fill a beaker with water and place it on a gauze on top of a tripod. Light a bunsen burner and place it under the tripod. Bring the water to the boil, then adjust the flame so that it goes on simmering gently.

2. Drop the leaf into the boiling water for about a minute (Fig 8.1(a)). This will make it soft and easy to penetrate by fluids. Then remove the leaf with a pair of tweezers.

3. *Turn out the bunsen burner.* Half fill a small beaker with ethanol and carefully place it inside the larger beaker of hot water. The ethanol will soon come to the boil.

4. Now drop the leaf into the beaker of ethanol (Fig 8.1(b)). After about five minutes the ethanol will remove the green pigment from the leaf, making it go a whitish colour.

5. Take the leaf out of the ethanol with tweezers, and wash it in tap water.

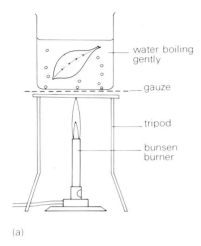

(a)

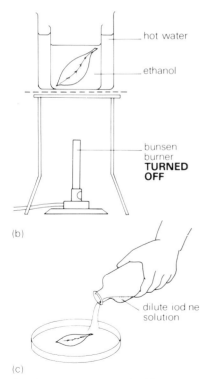

(b)

(c)

Fig 8.1 How to test a leaf for starch

6. Place the leaf in a dish and pour dilute iodine solution over it (Fig 8.1(c)). Watch what happens. If the leaf turns blue-black, starch is present.

Questions for class discussion

1. Did you find that the whole of your leaf turned blue-black? If it did not do so, can you suggest a reason?

2. Why was it necessary to turn out the bunsen burner before putting the beaker of ethanol into the boiling water?

3. When you were boiling the leaf in ethanol, what happened to the colour of the ethanol? Explain.

4. If you were to take a green leaf off a plant and immerse it in iodine solution, it would *not* go blue-black. Why not?

Light and starch-formation

Green plants grow only in places where there is *light*. What do they need light for? A possible reason is that they need light for making starch. We will now do an experiment to test this suggestion.

Experiment 8.2 To find out if a green plant needs light to make starch

Your teacher will give you two potted geranium plants which have had all the starch removed from their leaves. The starch has been removed by placing the two plants in darkness for several days.

1. Place the two de-starched plants in the window. Cover one of them with an upturned cardboard box so that it gets no light. Leave the other plant uncovered so that it gets plenty of light. Leave the plants for at least 24 hours.

2. After 24 hours, take a healthy green leaf from each plant and test it for starch with iodine (see Experiment 8.1, p. 108). Do you find that the leaf from the uncovered plant contains starch, whereas the leaf from the covered plant does not?

Do your results support the suggestion that the plant needs light for making starch?

Questions for class discussion

1. Why did you have to use de-starched plants for this experiment?

2. Why was it necessary to set up *two* plants, one in the dark and the other in the light?

3. It is important that both plants should be given exactly the same conditions except for the condition you are investigating, namely light. Why is this important?

The illuminated plant (the plant that was left in the light) in this experiment provides a standard with which the darkened plant may be compared. We call the illuminated plant the *control plant*. An experiment involving a control is called a *controlled experiment*. Controlled experiments are very important in biology, and you will meet them frequently.

Photosynthesis

We have seen that plants make starch in the light. This process is called *photosynthesis* (*photo* means 'light' and *synthesis* means 'making' or 'manufacturing'). Starch is the plant's food, and so the plant is able to *make* its own food. In order to do this, energy is required and the energy comes from sunlight.

Another thing that is needed for photosynthesis is the green pigment *chlorophyll* which is found in all green parts of the plant, particularly the leaves. You saw this pigment when you decolourised your leaf in Experiment 8.1.

Experiments show that the raw materials for photosynthesis are carbon dioxide and water, and that oxygen gas is given off as a by-product.

We can sum up photosynthesis like this:

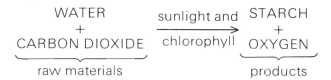

The words 'sunlight and chlorophyll' are written by the arrow because they are needed for photosynthesis to take place. The function of the chlorophyll is to absorb the energy from sunlight so that it can be used by the plant for making starch.

Starch is not formed straight away. Sugar is formed first, and this is then turned into starch. In plants such as sugar cane and sugar beet, the sugar which is made by photosynthesis is not turned into starch but remains as sugar. So these plants are used by us as a source of sugar (Fig 8.2).

Fig 8.2 Sugar cane: the thick stems contain sugar which has been made in the leaves by photosynthesis

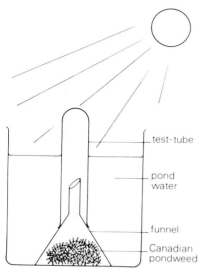

Fig 8.3 Experiment to find out if Canadian pondweed gives off oxygen in the light

Experiment 8.3   To show that a green plant produces oxygen

For this experiment we will use a water plant called Canadian pondweed (*Elodea canadensis*) which is a common green plant in British ponds and lakes.

1. Place a handful of Canadian pondweed under a funnel and set it up as shown in Fig 8.3. The upturned test-tube must be full of pond water.

2. Leave the pondweed in a well lit place for about a week.

3. Set up another sample of Canadian pondweed in exactly the same way, but leave it in the dark for a week.

4. After a week examine the two samples of pondweed. Has any gas gathered in either of the test-tubes? If so, in which one?

    You should find that only the illuminated pondweed has given off any gas; in fact you may see bubbles of gas rising from it.

How could you show that this gas contains oxygen? The simplest test would be to see if it will ignite a glowing splint (Fig 8.4).

Question for class discussion

Why was it necessary to have two samples of pondweed in this experiment, one in the light and one in the dark?

Homework assignments

1. Photosynthesis is a green plant's way of making food.
a. List four things which a leaf must have for photosynthesis to take place.
b. Which of these things are 'raw materials'?
c. Describe an experiment which you could do to show that a green plant cannot make starch if one of the four things which you have listed is missing.

2. One way of finding out if light is needed for a leaf to make starch is shown in Fig 8.5.
a. What should be done to the plant beforehand, and why?

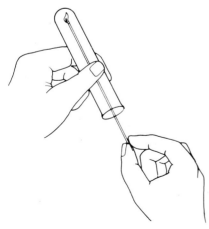

Fig 8.4 Testing the gas given off by Canadian pondweed for the presence of oxygen

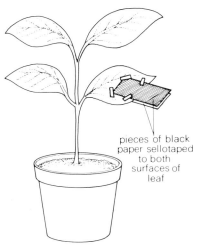

pieces of black paper sellotaped to both surfaces of leaf

Fig 8.5 Experiment to find out if light is needed for a leaf to make starch

b. If after 24 hours you test the leaf with iodine, what would you expect the result to look like?

c. Where is the control in this experiment?

d. Is the control satisfactory? If not, can you suggest a better one?

3. Your friend was away when you were shown how to test a leaf for starch. Write out some instructions for him, stressing any precautions he should take.

4. Suggest two reasons why photosynthesis is important to man.

5. In the 17th century a scientist called van Helmont did the following experiment. He planted a willow in a pot of soil, having weighed the willow and the soil separately beforehand. He then left the willow for five years, making sure that the soil was kept well watered. Five years later he dug up the willow and weighed it, and he also weighed the soil on its own again. His results are shown in Table 8.1.

Table 8.1

	Before planting	Five years later
Mass of willow	2 kg	77 kg
Mass of soil	100 kg	100 kg

How had the willow managed to increase in mass?

6. In the 18th century Joseph Priestley carried out the following experiment. He burned a candle in a sealed chamber until the flame went out. He then divided the air into two separate glass containers. In one container he placed a green plant; no plant was placed in the other container. Both containers were put in a sunny place. Ten days later he found a lighted candle would burn in the first container, but not in the second. How would you explain this?

Obtaining light for photosynthesis

The part of a plant in which photosynthesis mainly takes

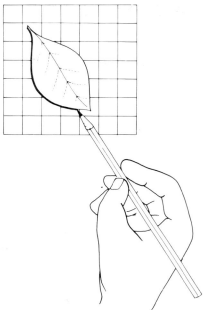

Fig 8.6 Tracing the outline of a
beech leaf on squared paper

place is the *leaves*. The cells inside the leaf contain chlorophyll. Leaves are generally flat, and together they have a large surface area for catching as much light as possible.

Experiment 8.4 Estimating the leaf surface of a tree

The best kind of tree to use for this experiment is beech which has simple leaves all more or less the same size.

1. Detach one leaf and lay it on a sheet of squared paper. Trace the outline of the leaf with a pencil (Fig 8.6).

2. From the squares that fall inside the outline of the leaf, work out the approximate surface area of the leaf in square centimetres.

3. Now estimate the approximate number of leaves on the tree. This is not an easy thing to do; if you do not know how to set about it, you may find Chapter 12 helpful (see sampling, p. 151).

4. Calculate the leaf surface of the tree by multiplying the surface area of a single leaf by the total (estimated) number of leaves. Give your answer in square metres.

5. Measure the dimensions of your laboratory and work out the floor area in square metres.

Questions for class discussion

1. How does the leaf area of your tree compare with the floor area of your laboratory? Why is it useful to make this kind of comparison?

2. How could you make your estimation of the leaf area more accurate?

3. Suppose you wanted to estimate the leaf area of a plant with only 25 leaves. How would you do it?

4. Plants generally have a large number of small leaves rather than a small number of large leaves. Suggest reasons for this.

Obtaining carbon dioxide for photosynthesis

The carbon dioxide which a plant needs for photosynthesis comes from the surrounding air.

Leaves are covered with a thin 'skin' called the *epidermis*. The epidermis has tiny holes in it called *stomata* (singular: *stoma*) (Fig 8.7). Carbon dioxide passes through the stomata to the inside of the leaf where photosynthesis takes place.

Experiment 8.5   Looking at stomata
The leaves of Ivy-leaved Toadflax are very good for looking at stomata because they are thin and semi-transparent. Ivy-leaved Toadflax can often be found growing on old walls.

Fig 8.7 Stomata seen in surface view down a microscope × 300

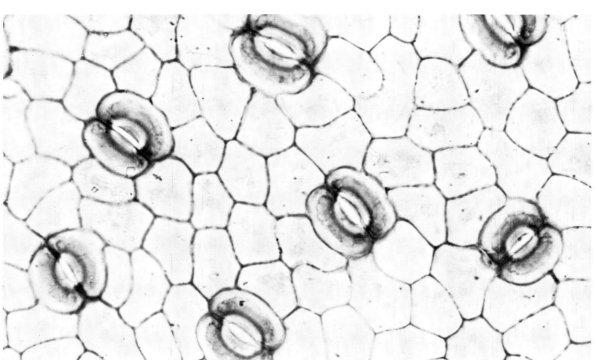

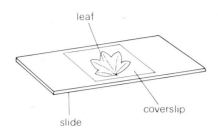

Fig 8.8 A leaf of Ivy-leaved Toadflax mounted on a slide

1. Remove a leaf from the plant and place it on a slide. *The lower side of the leaf should be uppermost.* Put a coverslip on top of the leaf so as to flatten it (Fig 8.8). There is no need to add any water or stain to the leaf.

2. Examine the leaf under the microscope, low power first, then high power.

3. Describe one of the stomata in detail, and make a sketch of it in your notebook.

Questions for class discussion

1. Could you see that the opening of the stoma is bounded by a pair of sausage-shaped cells? These are called *guard cells* (Fig 8.9). What do you think their function might be? Where are the guard cells in Fig 8.7?

2. What other things besides carbon dioxide might pass into, or out of, the leaf through the stomata?

3. How could you estimate the number of stomata in a square millimetre of leaf surface?

4. Most leaves have many more stomata on the lower side than on the upper side. Suggest a reason.

Obtaining water

The plant obtains its water from the soil: it is absorbed by the roots and taken up the stem to the leaves. Only a fraction of this water is used in photosynthesis. Most of it evaporates from the leaves through the stomata.

Experiment 8.6   Measuring the volume of water taken up by a plant

You will need a measuring cylinder with a capacity of 25 cm³.

1. Find a small plant growing near the laboratory. The above-ground part of the plant should have well developed leaves and should be about 10 cm high.

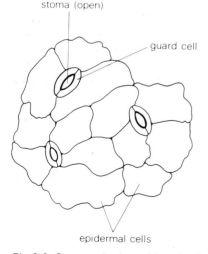

Fig 8.9 Stomata in the epidermis of a leaf of Ivy-leaved Toadflax

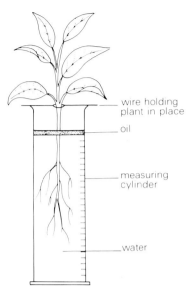

wire holding
plant in place

oil

measuring
cylinder

water

Fig 8.10 Apparatus for measuring the volume of water taken up by a leafy plant

Fig 8.11 The root of a young seedling showing the root hairs

2. With a trowel, dig up the plant very carefully so as not to damage the roots.

3. Take the plant into the laboratory and wash the soil off the roots *gently* with cold tap water.

4. Place the roots of your plant in the measuring cylinder. Then fill the measuring cylinder with water exactly to the 25 cm³ mark.

5. Carefully pour a little olive oil into the measuring cylinder, so that it forms a thin layer on the surface of the water. This should stop any water evaporating from the measuring cylinder (Fig 8.10).

6. Fill a second measuring cylinder with water up to the 25 cm³ mark, and pipette a little olive oil on to the surface as before. This will be your control.

7. After a week, record the level of the surface of the water in both measuring cylinders.

Questions for class discussion
1. In which measuring cylinder has the water level fallen most, and why?

2. Water does not always travel through a plant at the same speed; sometimes it travels very slowly but at other times it may travel much faster. The speed depends on how quickly water evaporates from the leaves. What conditions in the plant's environment are likely to speed up the rate at which water evaporates from the leaves?

3. A short way back from the tip, each root has a covering of very fine outgrowths called *root hairs* (Fig 8.11).
a. In what way might these root hairs help the roots to absorb water?
b. What other functions might be carried out by the root hairs?

Obtaining mineral salts

The roots also absorb mineral salts such as nitrates and

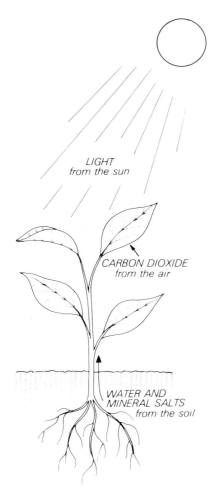

LIGHT
from the sun

CARBON DIOXIDE
from the air

WATER AND
MINERAL SALTS
from the soil

Fig 8.12 How a plant gets the things it needs for photosynthesis

phosphates. These are needed for the plant to make important substances such as proteins.

Mineral salts are present in soil, dissolved in the soil water. They come from the dead remains of animals and plants that have rotted (decayed) in the soil. Dead leaves, for example, provide a good source of mineral salts. Gardeners and farmers add *manure* or *fertilisers* to the soil to enrich it in mineral salts.

If a plant is deprived of any of the mineral salts that it needs, it will show poor growth and may die.

Figure 8.12 summarises how a green plant obtains the various things it needs for photosynthesis.

Gas exchange in a plant

Plants undergo two quite distinct processes, both involving an exchange of gases between the leaves and the surrounding air:

1. *Photosynthesis*, the plant's method of making food. In this process the plant takes in carbon dioxide and gives out oxygen. Photosynthesis requires light, and so it will only take place during the day. It stops at night.

2. *Respiration*, the plant's method of obtaining energy. In this process the plant takes in oxygen and gives out carbon dioxide. Respiration occurs all the time, at night as well as during the day.

How do plants affect the atmosphere?

During the day, when it is light, green plants take in more carbon dioxide for photosynthesis than they give out in respiration. The result is that they remove carbon dioxide from the air around them. This is particularly so around mid-day when the light is at its brightest.

At night, when it is dark, photosynthesis stops but the plants continue to respire. The result is that they now add carbon dioxide to the air around them.

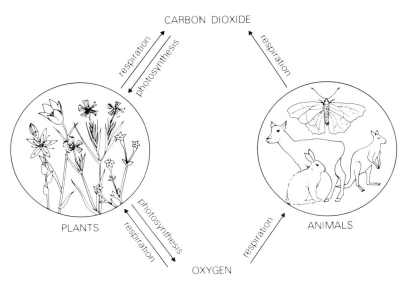

Fig 8.13 How animals and plants affect the atmosphere. Their combined activities keep the amount of carbon dioxide in the atmosphere more or less constant.

Animals, of course, respire all the time, so they constantly add carbon dioxide to the surrounding air.

Figure 8.13 sums up how animals and plants affect the atmosphere.

Homework assignments

1. Gardeners and farmers often add fertilisers to their soil. These fertilisers contain, amongst other things, the element nitrogen. Find out why plants need this element.

2. Scientists have found that, in an area with dense vegetation such as a wood, the amount of carbon dioxide in the air varies slightly.
a. At what time of day would you expect it to be at its highest?
b. At what time of day would you expect it to be at its lowest?
   Explain your answers.

3. At one time nurses used to take all the plants out of a hospital ward at night.
a. Why do you think this was done?
b. Suggest a reason why it is no longer thought to be necessary.

119

4. Scientists have made accurate measurements of the average amount of carbon dioxide in the air at different times of the year. They found that in the early spring the carbon dioxide content was very slightly greater than in the late summer. Explain the difference.

5. It is possible to grow plants such as mosses and small ferns in a stoppered jar that will not let in any fresh air. The plants will not grow much but they can last for years. How do they survive and why do they not grow much?

Summary

1. Green plants make starch by *photosynthesis*: the raw materials are carbon dioxide and water, and oxygen is a by-product. Sunlight provides energy for photosynthesis, and the green pigment *chlorophyll* is required.

2. The iodine test can be used to find out if a leaf contains starch. The leaf must be softened and decolourised first.

3. A controlled experiment can be done to show that light is needed for a green plant to make starch.

4. Another experiment can be done to show that Canadian pondweed will produce oxygen gas when illuminated.

5. Photosynthesis takes place mainly in the *leaves*. Green plants have a large leaf area for obtaining as much light as possible.

6. Water is absorbed from the soil by the *roots*. It then passes up the *stem* to the leaves from which it evaporates.

7. The roots also absorb mineral salts from the soil. These are needed for healthy plant growth.

8. The gas exchanges which take place in green plants can affect the amount of carbon dioxide in the surrounding air.

Chapter 9   # Sexual reproduction in flowering plants

How do animals reproduce sexually? You will remember that sexual reproduction involves two individuals, a male and a female. The male produces sperms and the female produces an egg. A sperm combines with the egg in the process that we call fertilisation. In flowering plants the same kind of thing takes place, though the details are different.

Fig 9.1 Daffodil flower cut open to show the structures inside

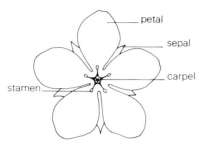

(a) The whole flower looking from above

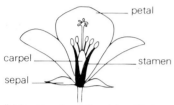

(b) Looking from the side with two sepals and two petals removed

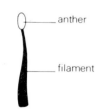

(c) A stamen, the male part of the flower

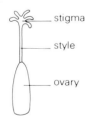

(d) A carpel, the female part of the flower

Fig 9.2 The structure of a geranium flower

The flower

The reproductive structures are found in the *flower* (Fig 9.1). There are many different kinds of flowers, but they all have certain features in common. We shall start by looking at a well known flower that shows these features.

Experiment 9.1   Looking at a geranium flower

Geraniums can be grown out of doors in the summer, and in pots indoors. You will need a plant that is in flower.

1. Detach a fully opened flower from the plant.

2. Observe the flower without pulling it to pieces.

You will see that it consists of a series of rings of structures (Fig 9.2(a)). From the outside inwards these structures are:

The *sepals*:   five small green leaves round the outside of the flower.

The *petals*:   five usually brightly coloured structures, larger than the sepals.

The *stamens*: five in all, each consisting of a stalk with a yellow knob at the end.

The *carpel*:   a tall, column-like structure in the centre of the flower with a swollen base and five short branches at the top.

3. Remove one of the sepals and examine it carefully under a hand lens. How does it differ from an ordinary leaf?

Look at an unopened flower bud on the geranium. Does this tell you anything about the function of the sepals?

4. Carefully remove two of the petals on one side of the flower. This will enable you to see the inside of the flower more easily (Fig 9.2(b)).

5. Examine a stamen in detail (Fig 9.2(c)). The yellow knob is called the *anther*, and the stalk is called the *filament*. Squeeze one of the anthers with a pair of tweezers. Can you see any small yellow specks? These are *pollen grains*. The stamens are the male part of the flower, and the pollen grains carry the male gametes which are equivalent to an animal's sperm.

Fig 9.3 How to slice open the carpel of a geranium flower

6. Observe a carpel in detail (Fig 9.2(d)). It consists of three parts: the *stigma* at the top, *style* in the middle, and *ovary* at the base. The ovary contains five little ball-like *ovules*, each of which contains a microscopic *egg*. The carpel is therefore the female part of the flower.

7. With a razor blade, slice open the ovary lengthways as shown in Fig 9.3. Can you see any ovules inside?

Questions for class discussion

1. Are the other flowers on the geranium plant identical with the one that you have studied? If not, how do they vary?

2. Not all plants have flowers similar to geranium flowers. From your memory of flowers which you have seen in your school grounds, or perhaps in your garden at home, how do the flowers of other plants differ from the geranium flower?

Experiment 9.2   Observing different kinds of flowers

Your teacher will give you the flowers of three different species of plant.

1. Examine each flower carefully and compare it with the others.

2. Write out Table 9.1 in your notebook.

Table 9.1

	Flower A	Flower B	Flower C
Name of plant			
Number of sepals Number of petals Number of stamens Number of carpels Colour of petals Shape of the flower Time of flowering (give months)			

3. Complete this table for your three flowers. Fill in as many of the spaces as possible from your own observations. Where necessary, look up the answer in a book.

Question for class discussion

List those features of the three flowers which are similar, and those features which vary from one species to another.

Homework assignments

1. The picture in Fig 9.4 shows a certain flower viewed from above:
a. Name the structures A, B and C.
b. Suppose you sliced through the middle of this flower along the dotted line; draw the probable appearance of the cut surface.

2. What part of a flower:
a. protects the flower bud before it opens;
b. contains an egg;
c. is equivalent to an animal's testis;
d. contains the male gametes?

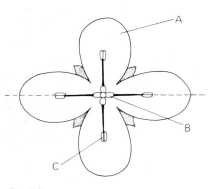

Fig 9.4

Pollination

For a flowering plant to reproduce sexually, the pollen grains must be transferred from an anther to a stigma. We call this process *pollination*.

In many plants, including the geranium, the pollen grains are transferred by insects such as bees, butterflies and moths (Fig 9.5).

When the insect visits a flower, it brushes against the anthers, and some of the ripe pollen grains stick to its hairy body. The pollen grains of some plants have little spikes projecting from them which make them 'sticky' and so help them cling to the insect. The insect then visits another flower, and some of the pollen grains fall off its body and stick to a stigma (Fig 9.6).

Why does the insect visit the flower in the first place? It does so in order to obtain *nectar*, a sugary fluid produced

Fig 9.5 A bee visiting a
Cotoneaster flower

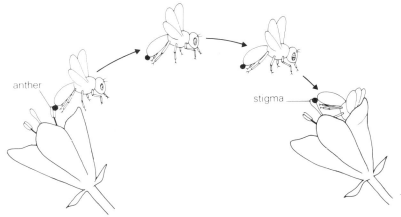

anther

stigma

Fig 9.6 These diagrams show how
a bee carries pollen from one flower
to another

at the base of the petals. Insects feed on nectar. The honey which bees make and store in their hive is concentrated nectar.

The flowers of many plants have special features which attract insects. For example, the petals may be large and brightly coloured, and their smell may be pleasant. Scientists have done experiments which show that insects are attracted by the colours and scents of flowers. What features of the geranium flower may attract insects?

Experiment 9.3   Watching insects visiting flowers

When you get a chance, watch insects visiting flowers. Record the name of the plant, and make a list of the insects which visit it.

Try to answer these questions as accurately as you can:

1. What part of the flower does the insect land on?

2. What does the insect do after it has landed?

3. How long does the insect spend visiting the flower?

4. Which features of the flower might attract the insect?

5. What features of the insect might help it to pick up pollen grains?

Pollination by wind

Pollen is not always transferred by insects. In some plants it is blown from one flower to another by wind.

An example of a wind-pollinated plant is hazel. The flowers are small and massed together into a catkin (Fig 9.7). The catkins hang down, and the pollen is readily scattered when they are shaken.

Experiment 9.4   Looking at pollen grains

You will need mature flowers from several different plants. The anthers should have split open to release their pollen grains.

Fig 9.7 Hazel catkins

With a paintbrush pick up a few pollen grains from an anther and place them on a microscope slide. Examine the pollen grains under the microscope.

Questions for class discussion

1. How could you estimate the approximate size of a pollen grain?

2. Does the appearance of the pollen grain suggest how it may be transferred from one flower to another?

3. Usually pollen grains are carried from the flowers of one plant to the flowers of a different plant of the same kind. We call this *cross-pollination*. However, on occasions a pollen grain may land on a stigma of the *same* flower. We call this *self-pollination*. Which do you think is better, cross-pollination or self-pollination? Explain your answer.

4. The pollen grains of wind-pollinated plants are usually smaller than those of insect-pollinated plants. Why?

5. The pollen grains of wind-pollinated plants are usually more numerous than those of insect-pollinated plants. Why?

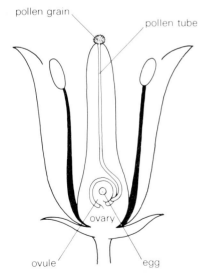

Fig 9.8 Fertilisation in a flower. The male nucleus is about to combine with the egg nucleus

Fertilisation

What happens when a pollen grain lands on a stigma? The pollen grain sends out a tube which grows into the stigma and down the style to the ovary. Eventually it reaches the egg. Then a male nucleus in the tip of the pollen tube combines with the egg nucleus (Fig 9.8). This is *fertilisation*. Do not confuse fertilisation with pollination which is simply the transfer of the pollen grains from the stamen to the stigma.

127

Experiment 9.5　Watching the growth of pollen tubes

1. Put a drop of sugar solution (10 per cent sucrose) on to the centre of a slide.

2. Obtain a flower with opened anthers. With a paintbrush pick up a few pollen grains from an anther and place them in the sugar solution. Cover with a coverslip.

3. Set up a second slide, but this time put the pollen grains in a drop of water rather than sugar solution.

4. Place your two slides in a closed petri dish and leave them in a warm, dark place for at least 30 minutes – longer may be necessary. Then look at the pollen grains under the microscope (low power). Have they changed in their appearance? Describe what has happened.

Questions for class discussion

1. Why did you have to put the two slides in a petri dish in this experiment?

2. Did all the pollen grains send out pollen tubes, or only some of them? If only some did, can you suggest a reason?

3. There might be a substance in the stigma which makes the pollen grain send out a pollen tube. Does this experiment help you to guess what this substance might be? What could you do to confirm your guess?

4. Describe an experiment which you could do to find
a. the best temperature
b. the best sugar concentration
   for the development of pollen tubes by the pollen grains of a given species of plant.

5. How do the male gametes of a flowering plant differ from the male gametes of a mammal such as the human?

Homework assignments

1. Study Fig 9.9, then answer the questions opposite.

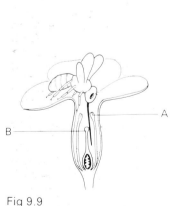

Fig 9.9

128

a. Name a substance which the bee obtains when it visits the flower.
b. How is the structure of the bee suited for obtaining this substance?
c. What does the bee use this substance for?
d. Name structures A and B.
e. What does the plant gain from being visited by bees? Explain your answer fully.
f. Bees are insects. Write down one feature of the bee, visible in the picture, that tells you it must be an insect.

2. Carefully explain the difference between:
a. self-pollination and cross-pollination,
b. pollination and fertilisation.

3. The picture below shows the pollen grain of a species of plant, greatly magnified. From the appearance of the pollen grain, how do you think the plant is pollinated?

Fig 9.10 A pollen grain magnified 1000 times

After fertilisation

When the egg has been fertilised, the fertilised egg (zygote) divides repeatedly and develops into an *embryo*. The embryo becomes surrounded by a hard wall. This hard wall, with the embryo inside, is known as the *seed*.

The seed contains a store of food, usually starch. This will nourish the embryo later when it develops into a new plant.

The final stage in the formation of a seed is that it gets extremely dry, and in this state the embryo becomes *dormant* – it goes to sleep, rather like an animal hibernating. However, it is still alive and capable of developing into a new plant. It may remain in this state for months or even years.

Fruits

The part of the flower surrounding the seed, or seeds, develops into the *fruit*. In most plants the fruit is formed from the ovary which, after fertilisation, may swell up considerably.

Experiment 9.6   Watching fruits forming

Some biological experiments take a long time, and this is one of them. Be patient!

Select a flower of a potted geranium plant which has been pollinated. Observe the flower every time you come into the laboratory. Make notes and sketches to show any changes which take place in its appearance.

What happens to the sepals, petals, stamens and carpel?

Which part of the flower develops into the fruit?

Examine a fully formed fruit, and make a drawing of it in your notebook (Fig 9.11).

Carry out similar observations on other flowers as instructed by your teacher.

Fig 9.11 The fruit of a geranium. It is known as the 'crane's bill' because of its shape

Dispersal of seeds

The job of the fruit is to help to carry the seeds as far as possible from the parent plant. We call this process *dispersal*. Why do you think it is important for the seeds to be dispersed as widely as possible?

Fruits help to disperse their seeds in four main ways:

1. *Some fruits are eaten by animals such as birds*. The seeds are not digested but pass out with the bird's droppings, often a long way from where the bird ate them. An example is cherry: the stone contains the seed.

   Fleshy fruits often look and taste nice, so animals are attracted to them.

2. *Some fruits disperse their seeds by splitting open*. This may occur with such force that the seeds are scattered quite a long way from the parent plant. Plants belonging to the bean family have fruits of this sort: the 'pod' is the fruit and the 'beans' are the seeds (Fig 9.12).

Fig 9.12 An opened bean pod

3. *Some fruits are covered with little hooks* which enable them to cling to the fur of animals, thus aiding the dispersal of the seeds. A well known example is the fruit of burdock which clings to your clothes (Fig 9.13).

4. *Some fruits have wings or hairs* which slow their fall, allowing them to be carried away by the wind. Sycamore trees have winged fruits, and dandelions have hairy parachutes (Fig 9.14).

Fig 9.13 Burdock fruits

Fig 9.14 Dandelion fruits

Experiment 9.7   Fruits and dispersal

You will need a fully formed fruit from a geranium plant.

1. Lay the fruit on a warm hotplate, or place it in a desiccator, to dry out. Watch carefully. Does it split open? If not, look at one which has already split open. Note its appearance (Fig 9.15). How does this fruit release and disperse its seeds?

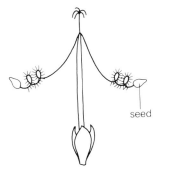

Fig 9.15 The geranium fruit after it has split open

Fig 9.16 The fruits of four well-known plants

(a) pea pod

(b) sycamore fruit showing 'wings'

(c) dandelion parachute

(d) fruit of goosegrass

2. Your teacher will give you the fruits of other kinds of flowers. Examine them carefully. In each case cut the fruit open and look for the seeds. How do you think the seeds are dispersed?

Questions for class discussion

1. The following are well known features of hedgerows in Britain: Hips, Haws, Sloes, and Old Man's Beard. What are they?

2. What tree do conkers come from, and what is the conker?

3. Figure 9.16 shows the fruits of four well known plants. Explain how each fruit aids in the dispersal of its seeds.

Homework assignments

1.
a. Why do most flowers come out in the spring and summer?
b. Why do most fruits appear in the autumn?

2. What happens to each of the following structures after the eggs inside a flower have been fertilised:
a. the petals;
b. the ovule;
c. the ovary;
d. the stamens?

3. A biologist wished to show that geranium flowers cannot make fruits unless they have been pollinated. He shut an unopened flower in a polythene bag, and looked at it carefully every day. A week later he was surprised to find that a fruit had appeared!
a. How might the flower have been pollinated?
b. What should the biologist have done to prevent pollination?
c. What should the control have been in his experiment?

4. Give the name of a flowering plant which you have studied.
a. In what sort of habitat is it normally found?

b. At what time of year does it flower?
c. How is its pollen transferred?
d. What features of the flower suit this method?
e. How does it disperse its seeds?

Summary

1. The part of a plant responsible for sexual reproduction is the *flower*.

2. The flower consists of *sepals*, *petals*, *stamens* and one or more *carpels*.

3. In the process of *pollination*, pollen grains are transferred from the anthers of one flower to the stigmas of another flower. Pollination may be brought about by insects or by wind.

4. *Fertilisation* involves the growth of a pollen tube into the carpel. A male nucleus in the pollen tube combines with the egg nucleus in the ovary.

5. After fertilisation the zygote develops into an *embryo*, the ovule into the *seed*, and the ovary into the *fruit*.

6. The seed has a protective coat and inside there is a store of food as well as the embryo.

7. Fruits have special features which help to disperse the seeds.

Chapter 10 **Growth and development of flowering plants**

In the last chapter we saw how sexual reproduction in flowering plants results in the formation of seeds. In this chapter we shall see what happens to the seeds.

Germination of seeds

After being dispersed, the seeds may remain in the soil through the winter. In the following spring or summer, each seed bursts open and a new plant starts growing out of it. We call this process *germination*.

Why is it important that germination should be delayed until the spring or summer? What do you think prevents germination taking place in the middle of winter?

What do seeds need to germinate?

What sorts of things do you think a seed *might* need in order to germinate? Make a list of them in your notebook. Think of an experiment which could be done to test each of your suggestions.

One of your suggestions is probably water. We will now do an experiment to find out if water is needed for germination.

Experiment 10.1 To find out if water is needed for germination
For this experiment use the seeds of mustard or cress.

1. Set up two dishes side by side. Place moist blotting paper in one of them, and dry blotting paper in the other.

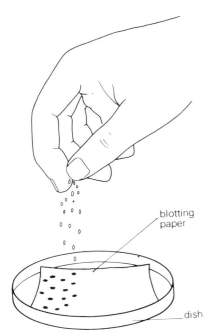

Fig 10.1 Sowing seeds of mustard or cress on blotting paper

2. Sprinkle the seeds on to the blotting paper in each dish (Fig 10.1). Cover the dishes, and place them in a warm, well lit place.

3. Observe the seeds at intervals during the next few days.

Questions for class discussion
1. Do your results support the suggestion that water is needed for seeds to germinate?

2. Why was it necessary to set up a dish with dry blotting paper as well as one with moist blotting paper?

3. In fact the following conditions have been found to be needed for the germination of seeds in temperate parts of the world such as Britain: water (moisture), oxygen and a temperature of between about 10°C and 30°C.

    Why do most seeds germinate in the spring or summer?

Homework assignments
1. A single weed may produce as many as one million seeds. However, only a small percentage of these germinate. Write down as many reasons as you can think of why the rest do not germinate.

2. Cress seeds were sown on moist blotting paper in three petri dishes and then placed as follows:
    A in an incubator at 50°C;
    B on a warm windowsill;
    C in a refrigerator at 4°C.
    Which dish of seeds would you expect to germinate first? Give two reasons for your choice.

Experiment 10.2   The absorption of water by a seed
You will need a dry broad bean seed for this experiment.

1. Weigh the dry seed, and write down its mass in grams.

2. Put the seed in a jar of water and leave it for about 24 hours.

3. Take the soaked seed out of the water and blot it gently. Then weigh it, and write down its mass.

4. Subtract the dry mass from the wet mass. This gives the increase in mass.

5. Express the increase in mass as a percentage of the original mass. We call this the percentage increase.

$$\text{Percentage increase in mass} = \frac{\text{increase in mass}}{\text{original mass}} \times 100$$

Questions for class discussion

1. Why was it necessary to blot the soaked seed before weighing it? Why did you have to blot it *gently*?

2. Why is it useful to express the increase in mass as a percentage of the original mass?

3. How do you think the seed takes up water?

4. Why does the seed need to take up water before it can germinate?

The seedling

The young plant which grows out of the seed is called a *seedling* (Fig 10.2). The seeding has two main parts: the *root* which grows down into the soil, and the *shoot* which grows up into the air. In time the shoot develops green leaves.

Experiment 10.3 Observing the development of a broad bean seedling

You will need a broad bean which has been soaked for 24 hours. The one from Experiment 10.2 will do.

1. Pour water into a glass jar to a depth of about 3 cm.

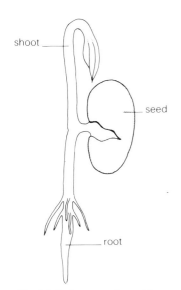

shoot

seed

root

Fig 10.2 A young broad bean seedling

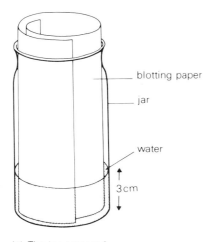

blotting paper

jar

water

3 cm

(a) The jar prepared

2. Roll up a sheet of blotting paper and place it in the jar as in Fig 10.3(a). The blotting paper should soak up the water and stick to the side of the jar.

3. Push the broad bean seed between the blotting paper and the side of the jar (Fig 10.3(b)).

4. Observe the jar at intervals during the next two weeks. Describe the changes which take place and record them in your notebook.

Questions for class discussion

1. When the seed began to germinate did you notice that the root appeared first, then the shoot? Why should the root appear first?

2. How many days after the experiment was set up did your seedling look like the one in Fig 10.2?

3. As the root grew out of the seed, did you notice that it was covered with fine hairs a short way behind the tip? What are these hairs for?

4. When the shoot grew out of the seed did you notice that the tip was bent back as in Fig 10.2? Suggest a reason for this.

How is the seedling nourished?

At first the young seedling gets all the food it needs from starch stored inside the seed. How could you show that a broad bean contained starch?

Later on the shoot develops green leaves, and the seedling starts making its own food by photosynthesis. How many days after germination did you observe green leaves on your seedling? The green colour is caused by the presence of a certain pigment in the leaf cells. What is this pigment called?

Homework assignment

A technician allowed a soaked broad bean seed to germinate

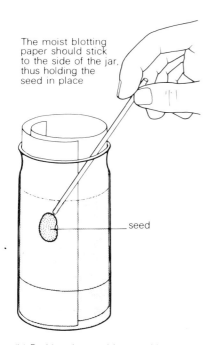

The moist blotting paper should stick to the side of the jar, thus holding the seed in place

seed

(b) Pushing the seed into position

Fig 10.3 Procedure for observing a broad bean seed germinating

in moist sand. She weighed the seedling every day between the third and twentieth days after germination. Her measurements are shown in Table 10.1

Table 10.1

Day after germination	Mass in grams
3	4.9
4	5.1
5	5.2
6	5.4
7	5.6
8	5.9
9	6.1
10	6.2
11	5.7
12	6.2
13	7.1
14	7.6
15	8.4
16	9.0
17	9.6
18	10.7
19	11.0
20	11.5

1. Plot these results on graph paper, putting days on the horizontal axis and mass on the vertical axis.

2. Explain the reason for:
a. the increase in mass between the third and tenth days;
b. the loss of mass between the tenth and eleventh days;
c. the increase in mass between the eleventh and twentieth days.

The importance of light to the developing seedling

When a seed germinates in the soil the shoot grows up and breaks through the surface of the soil. Once the shoot has broken through, light falls on it. Is light important for the further growth and development of the seedling? We can find out by growing seedlings in the dark.

Experiment 10.4   Investigating the effect of darkness on the development of seedlings

1. Sprinkle mustard or cress seeds in two dishes containing moist blotting paper.

2. Put one dish in a well lit place, and put the other dish under a cardboard box.

3. Observe the two dishes of seedlings at intervals during the next week and notice any differences in their appearance.

4. Make a careful drawing of a seedling from the well lit dish, and one from the darkened dish.

5. Make a list of the ways in which the seedlings grown in the dark differ from the well lit ones.

Questions for class discussion

1. A seedling which is grown in the dark differs from a well lit one in three main ways:
a. The stem is taller and thinner;
b. The leaves are yellow instead of green;
c. The leaves are smaller.
   A plant with this appearance is described as *etiolated*. What can you say about the functions of light in the development of a seedling?

2. What other things, besides light, are necessary for seedlings to grow and develop properly? Make a list of as many as you can think of.

Homework assignments

1. A boy took two similar broad bean seeds and planted them in two pots of soil. One (a) he left in a well lit place; the other (b) he placed in a dark cupboard. After two

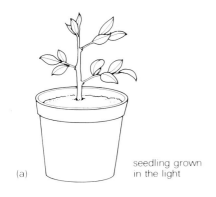

(a)   seedling grown in the light

(b)   seedling grown in the dark

Fig 10.4 Experiment to find out what happens to a plant left to grow in a dark place

weeks he examined them and found that they looked like the drawings in Fig 10.4. As well as the differences that can be seen in the drawings, plant (b) was yellow whereas plant (a) was green.

a. Write down two differences, other than colour, between the two plants.

b. Apart from the amount of light, name two other factors which might have produced the differences between these two plants.

c. Explain how you could carry out an experiment with broad bean seedlings to try to find out if it is only light that is producing the differences between these two types of seedlings.

2. Name a plant that you have studied. Describe briefly how you would measure the rate of growth of the plant.

b. Describe how you would carry out an experiment to discover whether this plant grows better when alone or in large groups.

Growth of the seedling into a new plant

It is fun to plant seeds and watch the young seedlings grow up into mature plants. At the same time you can co experiments to find out what sort of conditions give the best results. To do this you need to be able to measure the amount of growth which takes place in a given period of time.

Experiment 10.5   Measuring the growth of seedlings

1. Your teacher will give you a shallow box of soil containing young seedlings of barley, wheat or maize.

2. With a ruler measure the height in millimetres of each seedling (Fig 10.5), and work out the average height.

3. Repeat this at regular intervals for the next two weeks. The seedlings should be kept in a warm, well lit place and the soil should be kept moist by watering it regularly.

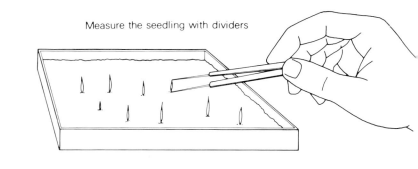

Measure the seedling with dividers

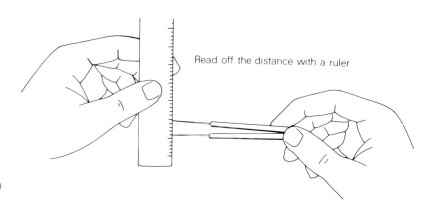

Read off the distance with a ruler

Fig 10.5 Measuring the height of a seedling

4. Plot your results on a sheet of graph paper: put the average heights on the vertical axis, and time in days on the horizontal axis.

Questions for class discussion
1. At the end of the two week period what is the height of
a. the tallest
b. the shortest plant?

2. How would you describe the shape of the curve on your graph? Explain the reason for the shape.

3. What is the height range in your sample (see p. 15)? Suggest possible reasons why some seedlings grow taller than others.

Fig 10.6 Sowing radish seeds in soil

Experiment 10.6 Growing radishes from seed

If it is summer you can grow radishes out of doors. Alternatively, fill a shallow box with moist soil almost to the top.

1. Make a channel in the soil about 2 cm deep. If you plant more than one row, the rows should be at least 30 cm apart.

2. Open a packet of radish seeds at the corner. Let the seeds trickle out into the channel in the soil (Fig 10.6).

3. Cover the seeds with a thin layer of soil. Pat the soil gently to make it firm. If you are using a soil box, leave the box in a warm, well lit place.

4. Examine the soil whenever you can. If necessary water it occasionally. How long does it take for the first shoots to appear?

5. Observe the plants over the next three weeks. Gently pull up one of the plants every two or three days and examine its roots; then throw it away.

Questions for class discussion

1. How long does it take for the roots to swell up into radishes?

2. The swellings are called root tubers and they are filled with stored food: this is why they are good to eat. What use are these tubers to the radish plant? Where does the stored food come from, and how does it get into the tubers? (Chapter 5 may help you to answer these questions.)

Homework assignments

1. A scientist planted a seed and as soon as the shoot appeared he measured its height. He continued to make measurements every day for three weeks. Conditions such as temperature and light were kept constant throughout this period. His measurements are shown in the graph in Fig 10.7.

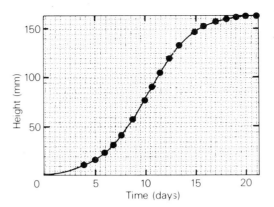

Fig 10.7

a. During which days was growth
   (i) fastest
   (ii) slowest?
b. Suggest possible reasons for the changes in the rate of growth which took place.

2. Two next door neighbours, Mr A and Mr B, shared a packet of onion seeds which they sowed in their back gardens. Mr A's onion plants turned out to be large and healthy, but Mr B's were small and unhealthy.
a. Suggest reasons why Mr A's plants were better than Mr B's.
b. Choose one of your suggestions and describe an experiment which you could do to find out if it is correct.

The life-cycle of flowering plants

The last two chapters have dealt with sexual reproduction, and the growth and development of flowering plants. Together these processes make up the life-cycle (Fig 10.8).

Some flowering plants grow to full size, produce seeds and die within one year. We call these *annual plants*. Other plants go on growing year after year, producing seeds each summer. We call these *perennial plants*.

Some examples of perennial plants are trees. Many trees shed their leaves in the autumn. They are known as *deci-*

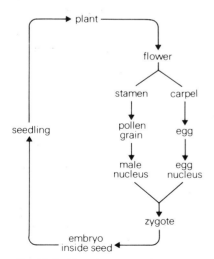

Fig 10.8 The life-cycle of a flowering plant

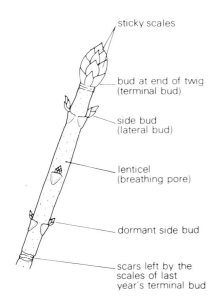

sticky scales

bud at end of twig
(terminal bud)

side bud
(lateral bud)

lenticel
(breathing pore)

dormant side bud

scars left by the
scales of last
year's terminal bud

Fig 10.9 A winter twig of a
horsechestnut tree

duous trees: examples are horsechestnut ('conker' tree) and beech. Other trees bear leaves all the year round. They are known as *evergreens*: examples are holly and yew.

Have you looked at a dec duous tree in winter? If you have you will have noticed that its twigs bear buds.

Experiment 10.7   What happens to the buds of a tree?

1. Cut off a winter twig of a horsechestnut tree about 20 cm from the end. Identify the parts of the twig shown in Fig 10.9.

2. Put the cut end of the twig in a jar of water. and place the jar in a warm, well lit part of the laboratory.

3. Observe the twig over the next few weeks. Pay particular attention to what happens to the bud at the end of the twig. List any changes in its appearance.

Questions for class discussion

1. Bringing a horsechestnut twig indoors causes the bud to open before it would have done out of doors. Suggest reasons why the bud opens more rapidly indoors.

2. What effect does the opening of the bud have on the length of the twig? How would this affect the overall size of the tree?

How do trees grow?

In Experiment 10.7 you saw how the winter buds on a tree produce new shoots in the spring. This has the effect of lengthening the branches.

At the same time the trunk gets thicker. If you look at the trunk of a felled tree you can see a series of rings (Fig 10.10). These are called *annual rings*, and they are new layers of wood which are produced each year. The outermost ring is the most recent layer of wood, and the innermost ring is the oldest. You can work out the age of a tree by counting its annual rings. How old do you think the tree in Fig 10.10 was when it was felled?

When is the new ring of wood formed each year? It happens when conditions are best for growth, which of course is mainly in the spring and summer. In the winter, when it is cold, very little growth takes place.

Fig 10.10 End-on view of the trunk of a felled tree showing the annual rings

Homework assignment

A boy measured the thickness of the annual rings in the outer region of the trunk of a felled tree, and he obtained the results shown in Table 10.2.

Ring number	Year of formation	Thickness
1	1977	5.0 mm
2	1976	1.0 mm
3	1975	8.0 mm
4	1974	5.0 mm
5	1973	1.5 mm
6	1972	4.5 mm
7	1971	5.5 mm
8	1970	4.0 mm

Table 10.2

a. Plot these results as a bar chart (histogram).
b. In 1975 there was an extra wide ring. What does that suggest about the growth of the tree?
c. What does it suggest about the weather conditions in 1975?
d. In which two years was growth slight?
e. Suggest *four* possible reasons for this slight growth.

Dormancy

Plants do not develop continuously all the time. At certain times of the year – usually the winter – the plant (or part of it) stops developing and goes into a prolonged sleep, rather like an animal hibernating. As explained in the last chapter, this is known as *dormancy*.

You have met two plant structures which become dormant in winter: seeds and buds. Dormancy is important because it enables the plant to survive an unfavourable period such as winter; it resumes its development when conditions become suitable.

Summary
1. *Germination* is the process by which a new plant starts growing from a seed.

2. To germinate successfully seeds require moisture (water), warmth and oxygen.

3. Before a seed can germinate it must absorb water, which greatly increases its mass.

4. When a seed germinates the *root* grows out first, and then the *shoot* appears. The young plant is called a *seedling*.

5. At first the seedling is nourished from food-stored inside the seed. Later the seedling develops green leaves which make food by photosynthesis.

6. For the seedling to grow properly, and for chlorophyll to develop in its leaves, light is needed.

7. Every spring the *buds* of a deciduous tree give rise to new leafy shoots.

8. A tree trunk increases in width by forming new layers of wood every year (*annual rings*).

9. Seeds and buds are *dormant* structures which enable the plant to survive unfavourable periods such as winter.

Chapter 11 # The numbers of living things

How many people can you see in Fig 11.1? You will probably say that it is impossible to count them. Biologists often need to know the number of each kind of animal or plant living in a habitat. This is called the size of the *population*. If we cannot count every individual then we have to guess how many there are; in other words we make an *estimate* of the population.

Fig 11.1 Competitors in the London Marathon

Experiment 11.1 How many organisms are there in an aquarium?

The aquarium which you set up in Chapter 1 should now contain large numbers of several species of organisms. In this experiment we shall estimate how many there are.

You will need a bench lamp, several shallow dishes, a hand net, wide-mouthed pipette and a hand lens.

1. Copy Table 11.1 into your notebook.

Table 11.1

Name of organism	Brief description	Population estimate

2. Examine your aquarium carefully. A bench lamp and a beaker will help you to look at the bottom of the tank (see Fig 11.2) but do not disturb the mud!

3. Start with the plants. Identify them as far as you can, using the key in the Appendix. If you cannot find the name, just write a brief description in your table.

4. Carefully remove one individual of each kind of small animal with a wide mouthed pipette or small net and place it in a shallow dish containing a little water. If it is

Fig 11.2 Examining the bottom of an aquarium through a beaker lowered into the water

too small to see clearly, use a hand lens to examine it. The key in the Appendix may help you to name it.

5. When you have identified most of the animals and plants, estimate the number of each species in the whole aquarium tank. Write your estimate in the third column of the table. It does not matter if your estimate is very approximate; it will still indicate whether the organism is common or rare.

If you have a pond, stream or canal near your school you could select a small area and estimate the size of the population of a species of animal or plant found there using this method.

Taking samples

A more accurate way of estimating the number of living organisms in a habitat is to choose a small area and count the number of organisms in it. Then work out how many small areas would fit into the whole habitat. The small area is called a *sample* area.

Taking samples will give you an accurate estimate of the size of a population if you *take as many samples as you can*. You must not choose your sample areas because they have a large number of individuals or because they are easy to count; the samples must be selected from all over the habitat, that is, *taken at random*.

Experiment 11.2   Estimating the number of duckweed plants in a pond

Duckweed is a plant which floats on the surface of ponds and canals. You may have some growing in your aquarium. Each plant consists of a single leaf (called a frond) with one or more roots growing out of it. Fig 11.3 shows the surface of a pond covered with duckweed.

1. The lines drawn on the photograph divide the surface of the pond into a *grid*. They are 1 cm apart. What is the area of one of the squares in the grid?

2. Draw the grid into your notebook.

Fig 11.3 Duckweed on a pond

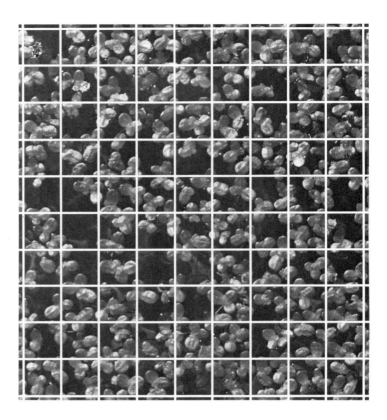

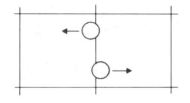

The arrows show the squares
in which each frond
should be included

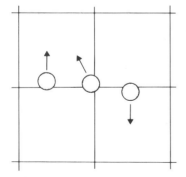

Fig 11.4 Counting duckweed
fronds

3. Count all the fronds in any *one* of the squares. Fronds which are on the edge of a square should be included if they are mainly in that square (see Fig 11.4). Write the number of fronds in the square in your notebook.

4. Count the fronds in nine more squares chosen randomly.

5. Write down the number of fronds in each square in your notebook, and add them up to find the total number of fronds in ten squares.

6. Work out the average number of fronds in one square. From this calculate the total number of fronds in the part of the pond visible in Fig 11.3.

Questions for class discussion

1. Use the sample method to estimate the number of animals in each of the photographs in Fig 11.5. Write down the number in each sample area as well as the total population in each picture.

Fig 11.5

(a) Gannet colony

(b) Herd of fur seals

(c) Flock of fieldfares

2. In Question 1, how many samples did you take to make your estimate of the number of animals in each photograph? Would it have been better to take (i) more, smaller samples or (ii) fewer, larger samples? Explain your answer.

Homework assignments

1. A pair of golden eagles requires a nesting territory of at least 2500 hectares. If the whole of the Cairngorm Mountains (25 000 hectares) is suitable for golden eagles, what is the largest number of pairs which could nest there?

2. The soil beneath a typical square metre of grassland may contain a hundred earthworms. How many worms are you running over when you are playing hockey? (A hockey pitch is approximately 4000 m².)

3. James and Mary decided to estimate the number of sticklebacks in a pond. Each of them caught sticklebacks at opposite ends of the pond. After 20 minutes James had caught 10 fish and Mary had caught 20. They concluded that sticklebacks were twice as common in Mary's half of the pond. Write down as many reasons as you can why that might *not* be true. Suggest how James and Mary could improve their method of estimating the number of sticklebacks in the pond.

4. You wish to find out how many bottles of the new drink 'Supa-Coola' should be taken on a school outing. You cannot ask all the pupils if they want one, so a sample must be asked. How would you choose your sample? From the results, how would you decide how many bottles to take?

How many plants are there in a lawn?

At first sight a lawn may appear to consist of nothing but grass. However if you look at it more closely you can see

Find a habitat with more than one species of plant distributed unevenly

Place a quadrat over a part of the habitat

Identify the species of plant present in the square

Imagine each species gathered together in one part of the square

Estimate the area occupied by each species and express it as a percentage of the square

30%	
20%	
50%	

Repeat for several squares chosen at random and average the results

Fig 11.7 Using a quadrat

Fig 11.6 A plant community in a lawn

that there are other plants as well, for example, clover, daisy and dandelion. Moreover there may be different kinds of grass.

All the different kinds of organism living together in a habitat make up a *community*; their positions in the community are called their *distribution*.

Look carefully at Fig 11.6 How many different species of plant can you see? Each species of plant prefers a different area. Clover is often found in hollows where the blades of the mower cannot cut it; chickweed is common on the higher parts of the lawn. We call this arrangement of species *uneven distribution*, and we study it by choosing small square sample areas called *quadrats*. A quadrat usually consists of a square wooden frame which is laid down on the ground (Fig 11.7).

Questions for class discussion

A group of pupils investigate the occurrence of a certain type of weed on a piece of waste ground. One of them places a quadrat frame on the ground. Figure 11.8 shows the positions of the weeds inside the frame.

1. What percentage of the area inside the frame is occupied by the weeds? How did you arrive at your answer?

2. Is your estimate accurate? How could you improve it?

3. When you use a quadrat frame to investigate the occurrence of plants on a piece of ground, you should put the frame on the ground *randomly* and in as many different places as possible. Why?

4. In some experiments you need to put the quadrat frame in definite positions on the ground rather than randomly. Can you think of an experiment where this would be necessary?

Fig 11.8 Diagram showing the position of weeds inside a quadrat frame

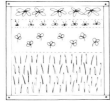

(a)
Place your quadrat frame flat on the ground as close to the tree, wall or path as you can.

(b)
Look carefully at the different kinds of plants inside the square and identify them.

(c)
Estimate the amount of the square which is covered by each species using the method described on page 155.
Write the percentage of each species in your table. This is sample 1.

(d)
Now use your metre rule to move the quadrat 1 metre away from your starting point. Estimate the percentages here and write them on the second line of your table (sample 2).

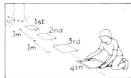

(e)
Take about 8–10 samples altogether, keeping to a straight line as far as possible.

Fig 11.9 Studying a lawn

Experiment 11.3 Studying the pattern of plants in a lawn

You will need a quadrat frame enclosing a square of 25 cm side. A metre rule will also be required. Work in small groups of 3 or 4. All groups should study the same habitat.

1. Your teacher will tell you which plants are present in the lawn you are studying and how to recognise them.

2. Copy Table 11.2 into your notebook.

Table 11.2

Sample	% grass	% plant A	% plant B . . . .
1			
2			
3			

3. With your notebook and pencil, quadrat frame and metre rule, take samples at the base of a tree, a wall or the edge of a path. Figure 11.9 (a) to (c) shows how to do this. Then take further samples (Figure 11.9 (d) and (e)).

4. When you return to the laboratory, make a bar graph of your results. Each member of your group should draw a graph for a different species of plant. Write 'sample number' along the horizontal axis and 'percentage of grass' (or one of the other species) up the side. Your graph might look something like this.

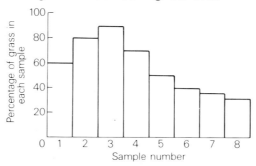

5. Write a heading above your bar graph and compare your results with those of other groups.

Questions for class discussion

1. Which quadrat sample in Experiment 11.3 contained the highest percentage of grass? Can you suggest a reason for this result?

2. Which quadrat samples contained the highest percentage of each of the other species you examined? Can you suggest why?

3. Compare your results with those of other groups. How much do they differ? Would your answers to questions 1 and 2 have been different if all the groups had added their results together?

Homework assignments

1. A group of boys and girls studied the distribution of grass and daisy plants around the base of a large pine tree in their school grounds. They also recorded the distribution of fallen pine needles. The results are shown in Table 11.3.

Table 11.3

Metres from base of tree	1	2	3	4	5	6	7	8	9	10	11
% pine needles	85	60	25	12	10	3	3	2	1	0	0
% daisies	0	0	0	0	0	1	6	8	10	15	30
% grass	15	40	75	88	90	96	91	90	89	85	70

a. What is the most likely explanation for the gradual decrease in the percentage of pine needles as you go further from the base of the tree?

b. Describe an experiment which you could do to test your explanation.

c. Why do you think there are no daisies close to the trunk? What could you do to find out if you are right?

d. The percentage of grass increases up to 6 metres from the trunk and then begins to decrease. Why do you think this is?

e. What results would you expect to get if you placed a quadrat 20 metres from the base of the tree? Write down percentage figures for each plant at this distance and explain briefly why you chose them.

2. Earthworms can be brought to the surface of the soil by spraying the soil with a 0.5 per cent solution of formaldehyde from a watering can. Describe how you would use this method to estimate the number of earthworms in a large field.

3. A friend makes his own wine by adding yeast cells to fruit juice. He ferments the wine in 5 litre jars (1 litre $= 1000 \, cm^3$) and checks the state of the population by counting the number of cells in a 0.1 cm³ sample every week. This week's count was 100 cells. How many yeast cells are there in the jar?

4. Gannets nest in colonies, usually on remote islands. Each pair of gannets defends a territory of roughly 1 m² around its nest. The largest colony in the world is situated on St Kilda, off the west coast of Scotland. It covers 10 hectares (1 hectare = 10 000 m²). How many pairs of gannets could nest in the colony?

5. Sitka spruce trees are planted 5 m apart in rows with 10 m between the rows. How many young trees would be needed to plant a 10 000 m² wood?

Populations can grow

Do you ever do the weeding in your garden? If you do, you will know how quickly weeds can appear again after you have done the weeding. Most populations of animals and plants can grow very rapidly. Their numbers increase in two ways:

1. New organisms move into the habitat (*immigration*). Every winter the resident birds in Britain are joined by thousands of birds immigrating from countries with colder weather conditions.

2. New organisms result from *reproduction*. Asexual reproduction (see Chapter 5) can produce particularly large numbers of offspring at a time.

The population game

Figure 11.10 is a program which can be used with a pocket calculator if it has a 'memory'. Discover how to enter figures in the memory; in some calculators, information can be stored in the memory by pressing the button marked 'Min' or 'STO'. Find out how information can be recalled from the memory; the button may be marked 'MR' or 'RCL'. Use the program to find out how rapidly a population of rabbits can increase.

1. Copy Table 11.4 into your notebooks.

Table 11.4

Generation	Time of year	Population
1	January	
2	April	
3	July	
4	October	
5	January	
6	April	
7	July	
8	October	

2. To start the game we will assume that a pair of rabbits produces a litter of ten babies each time they breed (i.e. in each *generation*). Press the buttons on the calculator in the order indicated in Figure 11.10.

3. Each time you reach the generation box, write down the number on the calculator display in your table. This gives the rabbit population at the end of each generation.

4. The rabbits breed every three months, so if you follow the instructions and go round the program eight times, you will find the population after eight generations, that is, two years.

Questions for class discussion
1. How many rabbits would the population contain after two years if each pair bred every three months and produced ten young in each generation?

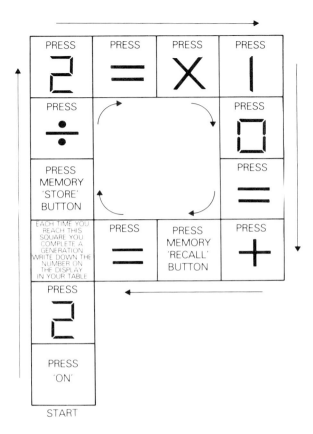

Fig 11.10 The population game

START

2. In fact the population is unlikely to reach this figure even if the rabbits did breed every three months and did produce ten young in each generation. Suggest two reasons for this.

3. Suggest two reasons why the rabbits' breeding rate might fall below the figures suggested above.

Homework assignment

A population of the duckweed *Lemna minor* was started with five fronds (see Experiment 11.2). The number of fronds was counted every three days. The results are shown in Table 11.5.

Table 11.5	Days from start of the experiment	Number of fronds	Days from start of the experiment	Number of fronds
	0	5	24	49
	3	7	27	65
	6	9	30	85
	9	12	33	100
	12	16	36	120
	15	21	39	125
	18	28	42	121
	21	37		

a. Plot these results on a graph, putting time along the horizontal axis and number of fronds on the vertical axis. Draw a smooth curve as close to the points as you can.
b. After the first two weeks the experimenters expected the final number of fronds to be more than 250. Why do you think they expected this? Suggest two reasons why the numbers of fronds did not increase as rapidly as they expected between two and six weeks.
c. Choose one of the reasons you have suggested and say how you could carry out an investigation to test it.

How populations are kept steady

You may have been surprised to find that two rabbits can produce half a million offspring in two years. But plagues of rabbits are rare. Something must prevent the rabbit population growing in this way. Figure 11.11 shows one of the things which kills rabbits. The disease *myxomatosis* first appeared in Britain in 1953 and is now an important check on the number of rabbits. As a result the rabbit population stays more or less steady.

Most animals and plants maintain steady populations under natural conditions. However, one animal with a population that has been growing for several hundred years is the *human* animal.

Human population growth

The total number of human beings in the world (*human*

Fig 11.11 Rabbit with
myxomatosis

population) was about 150 million in the year A.D. 1. The
changes since then are shown in Table 11.6.

Year (A.D.)	Estimated world population (millions)
1	150
1500	400
1650	550
1750	750
1800	900
1900	1600
1950	2400
1975	4000

Table 11.6

1. Plot a graph showing the growth of world population.
   Put the years (0–2000 A.D.) along the horizontal axis
   and the population (0–4000 million) on the vertical axis.

2. Join the points with a smooth curve.

3. What would you expect the population to be in the year 2000? Explain how you estimated your answer.

4. List as many things as you can which might slow down the rise in world population in the future.

The causes and results of human population growth

The remarkable increase in the human population during the last 150 years has not affected all groups of human beings. A few tribes in less developed parts of the world still show the same steady populations as most other species of animals and plants. This suggests that in communities in the more developed countries some of the factors controlling human population no longer work. Here are two possible explanations:

1. Improved food supply reducing deaths from starvation during periods of drought etc.

2. Improved medical care leading to fewer people dying of diseases.

Can you think of any other explanations?

The harmful effects of the increase in human population can be seen all round us. The environment is damaged by the waste materials that we release (*pollution*). As populations increase in size, the pressure on housing increases and living standards fall (Fig 11.12). Famine and natural disasters remind us that other countries suffer more than we do.

Homework assignments
1. How would you estimate the numbers of a small invertebrate living in a habitat such as a pond, small wood or stretch of seashore?

2. Explain the difference between the words *population* and *community* by using each word in a sentence describing a habitat you know.

Fig 11.12 Poor living conditions

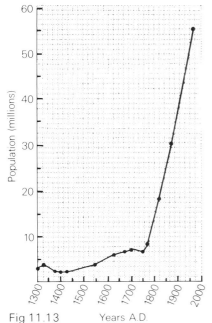

Fig 11.13　　Years A.D.

3. In the autumn a maple tree in a garden had 60 small maple seedlings growing beneath it. In spring all the seedlings except one had disappeared. The one remaining seedling was 20 metres from the base of the tree, further than any of the other seedlings. Why do you think this one seedling survived, and what caused the others to disappear?

4. Look carefully at the graph in Fig 11.13, which shows the changes in the population of Great Britain since 1300.
a. What was the population in (i) 1550 and (ii) 1950?
b. The 'Black Death' killed at least one third of the population of Britain. When do you think it occurred?
c. How long did it take for the population to return to the number which existed before Black Death?
d. During which 100-year period did the population increase by the greatest number?
e. The number of babies born per year in Great Britain has been falling recently, but the population is still rising. Can you suggest two reasons?

165

Background reading

Man against the pests

It has been estimated that about one third of the world's food crops are destroyed by a wide selection of pests before they are harvested. More are spoiled by pests infesting stored food. The most destructive of these pests is, without a doubt, the locust.

Locusts breed in very large numbers when the conditions are right. A high rainfall is needed just when the locust eggs are hatching. If the rain continues, more and more eggs will be laid, and swarms of hoppers, – the early stage in the locust life-cycle – will be produced. When they mature, they fly off in search of more food, and thus spread over wide areas. Swarms can cover as much as 500 km². A locust eats its own weight (2–3g) in food each day and a swarm covering 1 km² could contain 40 to 80 million locusts. This means hundreds of thousands of tonnes of vegetation are being devoured each day.

A huge area of the world, including 60 countries in North Africa and West Asia, have been threatened by the recent increase in locust numbers. This is a fifth of the earth's land surface and contains a tenth of the world's population. To control the billions of locusts it is necessary to spray the swarms, preferably at the hopper stage, and destroy at least 90% of the insects. But there are problems. One of the most common, cheap methods of control is spraying with DDT, and in some areas, previously untroublesome species are now multiplying as their natural enemies are wiped out by DDT. DDT <u>residues</u> are known to build up in animal tissues, and many birds and animals have declined in numbers because of this. Other forms of pest control need to be developed, such as the use of natural predators and safer chemicals. But a careful balance must be kept in the meantime between the destructive powers of pests and the side effects of control methods.

(Adapted from *Together for Children*, published by Oxfam/Unicef.)

Questions

1. Write down the meaning of the word 'residues'.

2. Explain in your own words how 'untroublesome species' might multiply after DDT treatment. Why might this be harmful?

3. Suggest ways in which 'natural predators' might help in the fight against locusts.

4. Two populations are mentioned in this passage. What are they? How do they affect each other?

Summary

1. The number of each kind of animal or plant in a habitat is called the size of the *population*.

2. All the different kinds of organism living together in a habitat make up a *community*.

3. The way in which living organisms are arranged in a community is called their *distribution*.

4. The number of living organisms in a sample can be *estimated* by taking *samples*.

5. The distribution of living organisms in a habitat can be studied by using small areas of the habitat called *quadrats*.

6. Populations can increase in number by *immigration* and *reproduction*.

7. Most natural populations of animals and plants are kept steady by a variety of checks such as disease.

8. Human populations have increased enormously in the last 200 years and have doubled in the last fifty years.

Chapter 12 # The distribution of living things

The lawn which you studied in Chapter 11 contained several kinds of weeds among the grass. You probably found that there were more weeds under trees and beside paths and that there was more grass in other parts. It is unlikely that any of the weeds were spread evenly over the whole area of grass.

The animals and plants in most habitats are unevenly distributed. In this chapter you will discover some of the reasons for this uneven distribution.

Experiment 12.1   The distribution of small invertebrates

In this experiment you will make your own collection of small animals by setting *pitfall traps*. The trap, which can be made from any small containers (for example, two plastic yoghurt pots), is dug into the ground so that the top of the pots is level with the surface of the soil (see Fig 12.1). Small animals walking along the ground fall into the container and cannot escape.

1. Draw a habitat map (see Experiment 1.4 page 8) of the area you are going to use.

2. Try to set at least one trap in each different part of the habitat – under a hedge, in a flower border, under a tree and in a grassy area. Fig 12.1 shows how to set a pitfall trap.

3. Indicate the positions of your traps on the habitat map and copy Table 12.1 into your notebook.

4. Visit the traps as often as you can. Remove the inner pot and empty the contents into a large jar. Replace the trap carefully.

Table 12.1

Date	Time	Weather conditions	Trap position	Animals caught

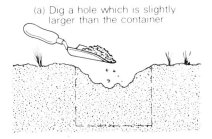

(a) Dig a hole which is slightly larger than the container

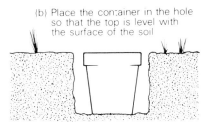

(b) Place the container in the hole so that the top is level with the surface of the soil

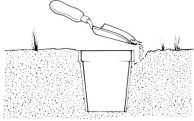

(c) Put a second container inside the first one. Fill in any gaps around the outer container.

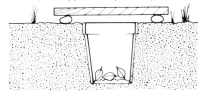

(d) Place a few leaves and small sticks in the trap to shelter the animals, and cover the top with a piece of wood supported by small pebbles

Fig 12.1 Setting a pitfall trap

5. Identify the animals using the key in the Appendix, and copy and fill in Table 12.1. Write down the number of each kind of animal caught in the trap in the last column, for example, woodlouse (3).

6. Return the animals to the places where they were caught.

7. After several days, gather together all the results. Find the total number of each kind of animal caught in each trap.

Questions for class discussion

1. In the questions which follow, give actual numbers *and* an explanation.
a. Which trap caught most animals? Which trap caught the least?
b. Which species was most frequently caught?
c. Were more individuals caught in the traps during the day-time or overnight?
d. Which species was most common in the day-time and which one at night?
e. Was there any species which was only caught in one area or in one trap?

2. Do your results show that the animals are unevenly distributed? Select one species with a distribution that is clearly uneven and indicate the areas where it lives on your habitat map. Is there a connection between the distribution of this species and the distribution of plants?

3. Did you catch more or less animals in the traps during wet weather? Can you suggest why rainfall might affect the distribution of small animals?

169

Homework assignment

1. Look carefully at the photograph in Fig 12.2. The plants are unevenly distributed.
a. Describe the experiments you would perform to record the distribution of the plants.
b. Suggest an explanation for their distribution.
c. Describe an experiment which you could do to test your explanation.

Fig 12.2

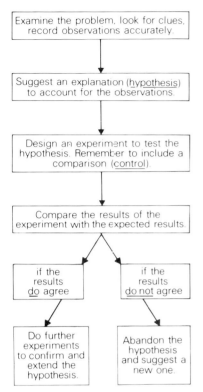

Examine the problem, look for clues, record observations accurately.

↓

Suggest an explanation (hypothesis) to account for the observations.

↓

Design an experiment to test the hypothesis. Remember to include a comparison (control).

↓

Compare the results of the experiment with the expected results.

if the results do agree | if the results do not agree

Do further experiments to confirm and extend the hypothesis. | Abandon the hypothesis and suggest a new one.

Fig 12.3 The way a biologist works

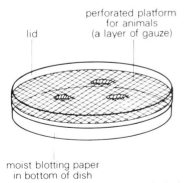

lid

perforated platform for animals (a layer of gauze)

moist blotting paper in bottom of dish

Fig 12.4 A common type of choice chamber

Looking for explanations

Biologists often have to work like detectives. They need to observe carefully and be on the lookout for clues. An uneven distribution of plants and animals in a habitat can be a useful clue. However, biologists have one great advantage over detectives: they can test their suspicions by doing experiments. Fig 12.3 shows what they do.

Now use this procedure in experiment 12.2.

Experiment 12.2   Choosing a place to live

If you were given ten minutes to collect as many woodlice as you could, where would you look? Why would you expect to find them there? What conditions do you think they prefer?

Your answer will probably include darkness as one of the things that woodlice prefer. One way to test this hypothesis is to use a *choice chamber*. There are several different kinds, but they all work in a similar way (see Fig 12.4).

To test the hypothesis that woodlice prefer dark places, follow the instructions given below.

1. Collect 10–12 woodlice and place them in a dark, moist container for ten minutes.

2. Open the choice chamber, remove the perforated platform and place moist blotting paper or cotton wool at the bottom of the chamber.

3. Replace the perforated platform, making sure that it is tight and smooth all round.

4. Fit the lid and leave the choice chamber for a few minutes so that a moist atmosphere builds up inside.

5. Take the lid off quickly and place about ten animals inside the chamber, spreading them evenly over the gauze. Replace the lid.

6. Cover one half of the chamber with black material and leave the other half uncovered.

7. After two minutes, record the number of animals on the dark and light sides.

8. Spread the animals out again and then cover the other half of the chamber.

9. After two minutes, record the numbers again. Keep changing the light and dark sides every two minutes until it is clear which side the woodlice prefer.

10. Now cover the *whole* container and record the positions of the animals after two minutes by means of a simple sketch.

11. Finally uncover the whole container and record the positions after two minutes in the light. Watch the animals this time and notice how and where they move before settling down. Make sure you do not cast a shadow on the container and that the light is even.

Questions for class discussion

1. Why were the woodlice placed in a dark, moist container for ten minutes before the experiment (instruction 1)?

2. Why was it important to leave the choice chamber so that a moist atmosphere built up inside it (instruction 4)?

3. Why were the dark and light sides of the choice chamber interchanged every two minutes (instruction 9)?

4. Why was the whole container covered (instruction 10)? What is the name given to this part of the experiment?

5. Briefly describe your observations when the container was left uncovered, and suggest an explanation for the final distribution of the woodlice (instruction 11).

6. Explain why the habit of avoiding light is useful to woodlice.

7. In your experiment, one side of the choice chamber was light and the other side was dark. Can you suggest one other difference between the two sides which might have affected the behaviour of the woodlice?

Homework assignments

1. Design an experiment which would test the hypothesis that 'woodlice prefer damp places'. Remember to include suitable controls.

2. The distinguished scientist Imek Broster has recently described the results of his experiments on the elusive millilouse, a small crustacean found only on one mountainside in Wales. Millilice can return across several miles of barren hillside to their own particular stone in total darkness. Broster explains their powers of navigation by suggesting that they use a built-in compass to find their way.

a. What is the problem that Broster is trying to explain?
b. Write down his hypothesis in your own words.
c. Describe an experiment which he might have performed to test his hypothesis, and say what results he would have obtained.
d. Suggest an alternative hypothesis to explain the navigation of millilice.

Reaching new habitats

Figure 12.5 shows the island of Surtsey which lies off the south coast of Iceland. This island is remarkable because it has only existed since 1965 when a volcanic eruption on the sea bed produced a pile of rocks and larva which cooled to form Surtsey. Biologists visiting the island a few years later found that a number of living organisms – mostly plants and insects – had reached it and started to live there.

The arrival of an organism in a new habitat is the result of *dispersal*. When the organism has established itself, we say it has *colonised* the habitat. If an animal or plant is absent from a habitat it may be because it has not reached it, or because it has failed to colonise the habitat.

Question for class discussion

The island of Krakatoa in the East Indies erupted in 1883 leaving a lifeless surface, 40 kilometres away from the nearest island. Fifty years later, 47 species of vertebrates had

Fig 12.5 The island of Surtsey

colonised the island including 36 birds, 5 lizards, 3 bats, a rat, a crocodile and a python. How do you think each of these groups of vertebrates reached the island?

Experiment 12.3   Colonisers in your aquarium

If your aquarium has been kept outside the laboratory, it will probably now contain several species of animals and plants which were not included when you originally stocked it up. Compare your lists of species from Experiment 11.1 with Experiment 1.3 and write in your notebook the names of all new arrivals.

The amount of colonisation which has taken place in the aquarium can also be studied by examining the microscope slides which have been suspended in the water.

1. Wipe the *lower* surface of the slide clean using a paper handkerchief or towel and place it on the microscope stage.

2. Using low power, examine as much of the slide as you can, looking carefully at all the living organisms.

3. Count the number of different kinds of animals and plants on each slide, and write the numbers in your notebook.

4. If you have time, draw and identify the species present.

5. Alternatively you could scrape the side of the tank and place the scrapings in a drop of tap water on a slide.

Questions for class discussion

1. How many new species of animals and plants have appeared in your aquarium? By what means might they have arrived there?

2. Two similar aquarium tanks were stocked with the same species of animals and plants. One was left outside while the other was kept in the laboratory under bench lights. A year later the two tanks were compared and the one which had been kept outside contained twice as many species as the other. Suggest *two* reasons for this difference.

Homework assignments

1. Mary set up an aquarium in September and hung twelve microscope slides in the water. Each month she removed a slide and examined it. At first the number of species increased steadily from month to month, but in spring and summer the number began to fall again. Suggest an explanation for
   a. the steady rise
   b. the gradual fall in the number of species.

2. Having baked a quantity of soil to kill any organisms it might contain, a man placed the soil in a plant tub outside his house. After a year he noticed the following

plants growing in it:

seedlings of yew, holly, oak and sycamore; a patch of moss;

some mould growing on the side of the tub where it was kept wet by water dripping from the roof above;

small flowers of shepherd's purse, thistle and groundsel.

Choose any four of these plants and explain how they might have reached the soil in the tub without any human help.

Becoming established in a habitat

What does an organism need in order to survive and multiply in a habitat? How many different things can you think of? List them in your notebook.

Even if all these requirements are available, the animals and plants may not colonise the habitat. There may be other organisms already established which eat the same food or use up the oxygen. We call these *competitors*. There may also be organisms which feed on the newcomer. These are called *predators*. Competitors and predators keep out colonisers unless they can reproduce very rapidly.

The colonisation game

To play this game divide the class into groups of four, each group having a pack of playing cards. In the game we assume that an animal or plant has reached a new habitat. The cards represent this habitat. The aim of the game is to find out which group produces the greatest number of colonisers.

1. Take the playing cards and remove the king, queen and knave of each suit. They will not be needed.

2. Choose one member of the group to be the first dealer.

3. The dealer invents the name of an organism and writes it down.

4. Shuffle the remaining cards thoroughly and place them in a pile, face down, in front of the dealer.

5. The dealer turns them over, one at a time, and gives the cards of each suit to a different member of the group; diamonds to one, hearts to another, spades to another and clubs to himself. They place the cards, face up, in front of them.

6. Each suit represents a different aspect of the habitat.
   Diamonds represent *water*
   Hearts represent *oxygen*
   Spades represent *food*
   Clubs represent *competitors and predators*

7. The aim is to collect enough water, oxygen and food before the organism is killed by competitors and predators. Each member of the group counts the total number of 'units' of water, oxygen etc. by adding up the 'pips' on the cards as they are turned over. When the total collected by a member of the group reaches 25, he informs the dealer.

8. If the totals of diamonds, hearts and spades *all* reach 25 before the total of clubs reaches 25, then the coloniser has survived. Go on to 9 to see whether successful breeding takes place. If not, change dealers, invent a new organism, shuffle the cards well and start again.

9. The dealer must now turn over an ace (of any suit) to represent successful breeding of his organism. If further clubs are turned, bringing the total to more than 25 before an ace is produced, then the organism has been eliminated before it could breed. Start again.

10. If the coloniser breeds successfully, the 'pips' on the cards which remain in the dealer's hand are added together to give the number of offspring produced by the coloniser. The dealer writes down his 'score', and a new dealer takes over.

11. The winner is the dealer with the greatest number of surviving and successfully breeding organisms when play ends.

Extreme environments

There are very few parts of the Earth which remain uncolonised by living organisms. Can you think of any?

Animals and plants may be specially suited or *adapted* to survive in places where other species would find it difficult to live. Here are some examples.

Polar regions

The Arctic and Antarctic regions have average temperatures of 45°C *below* freezing point (see Fig 12.6). Animals such as penguins and polar bears have thick coats of feathers or fur, short limbs and squat bodies to reduce their heat loss. This helps them to keep their temperatures constant. Other animals such as fish may contain 'anti-freeze' to stop their blood freezing.

Fig 12.6 A polar region
Cape Sherrard, Devon Islands

Fig 12.7 A desert region
Sahara Desert, Dcuz,
Tunisia

Desert regions

The Sahara desert has an average temperature of 35°C *above* freezing point and it rains, on average, once every three years (see Fig 12.7). Most desert animals keep out of the sun during the day if they can, and use very little water for excretion. Many animals and plants store food and water; plants may survive as seeds for a very long time until the rain comes. They then flower and produce more seeds within a few days. A few drought enduring plants have thick outer surfaces and small leaves to reduce evaporation, and long roots for absorbing water from deep down in the soil.

Mountain tops

Few living things are found at the tops of high mountains because of the combination of low temperature, strong

Fig 12.8 A mountain top habitat
Snowdonia, North Wales

winds, lack of water and lack of oxygen (see Fig 12.8). Few species are able to survive these all the year round. Most animals migrate to lower ground in winter. The plants are only a few centimetres high, thereby avoiding the wind and reducing water loss.

Homework assignments
1.
a. What is meant by the word habitat?
b. Name a habitat you have studied and describe its main characteristics.
c. Name the most common plant species in the habitat.
d. Suggest reasons why it grows there successfully.
e. How might this plant have arrived in the habitat?
f. How did it become established in the habitat?
g. Describe the way in which the plant survives the winter.

2. Living things do not normally inhabit totally dry places.
a. Give reasons why organisms need water.
b. Describe ways in which animals reduce their water loss.
c. Describe ways in which plants reduce their water loss.

3. A path runs through a grassy field. Five sample areas were chosen on the path, and five on each side of the path in the grassy field. Table 12.2 shows with a tick (√) the species present in each sample area.

Table 12.2

Species	Grassy Field					Path					Grassy Field				
	1	2	3	4	5	1	2	3	4	5	1	2	3	4	5
Grass type A	√	√	√	√	√							√	√	√	√
Grass type B		√	√	√	√						√	√	√	√	√
Grass type C						√	√	√	√	√		√			
Daisy					√	√	√	√	√	√		√			
Plantain						√	√	√	√	√		√			
Rock rose		√	√	√										√	√
Hawkweed			√		√	√		√	√	√	√				
Ribwort	√	√		√	√		√					√		√	√
Salad burnet	√	√	√	√		√			√		√	√		√	√

a. (i) Which species of plant is not found on the path?
   (ii) Suggest a reason why it is not found there.
b. (i) Which species is found only on the path?
   (ii) Suggest a reason why it is found there.
c. Which species grows most abundantly in both path and field?
d. Calculate the average number of species per sample area (i) in the grassy field   (ii) on the path.
e. Which of the following grasses is more likely to have been grass type C? Explain your choice.
   Poa annua Height 10 cm. Broad, blunt leaves.
   Brachypodium pinnatum Height 30 cm. Narrow, stiff leaves.

Background reading

Fish-life in the deep sea
In the early days of marine biology it was believed that below the sunlit zones of the sea there existed a lifeless

zone. However, a century ago, during the voyage of HMS *Challenger*, this theory was disproved as animals, most of them new to science, were dredged and trawled from the deep sea.

There is something fascinating in the capture of fishes in the great depths of the ocean. The deepest known capture is of a brotulid fish (*Bassogigas*) in the Puerto Rico Trench at a depth of over 25,000 ft, nearly five miles. Fishes of this genus have been found in several of the great trenches of both the Atlantic and Indian Oceans, and clearly they have specialised in deep sea life. Surprisingly in some ways they have a large and functional swimbladder, and it has been suggested that the oxygen stored in it may be used for respiration in areas where dissolved oxygen is low. If this is so it would be yet another example of the fascinating adaptations by fishes for life in difficult circumstances.

In the deep sea the density of animal life is lower than in the shore or surface regions of the sea and many of the adaptations are in response to this sparsity of life. In the absence of plant life all animals are predatory, and most of the deep sea fishes have huge teeth and enormous jaws. It seems that opportunities to feed are few, and no chance can be missed even if the prey is nearly as big as its captor, and occasionally fish are caught with prey in their gut as long as themselves.

The viper fishes (genus *Chauliodus*) which are found in all the deep seas are a perfect example. The fangs are so huge that the jaws cannot be shut tight, but the latter are loosely hinged so that they can swing wide open. It has a long curved ray with a lighted tip at the front of its dorsal fin which is dangled enticingly before its snout to attract prey. Many other deep sea fish have luminous lures on the snout which serve the same function.

Angler fish of the family Certiidae have an even more fascinating adaptation to the vastness of the deep sea. The females are often large (some species grow to 24 inches in length), but once the male attains an inch or so it attaches itself to the female and becomes parasitic upon her. The blood supply of the two fishes is joined and the male is supplied with such oxygen and nourishment as are necessary. When spawning takes place there is no question of

Fig 12.9 Angler fish

failure to find a member of the opposite sex; fertilisation of the eggs is ensured by the involuntary presence of the male.

Ceratoid angler fishes also possess other interesting features. Free swimming males have larger nasal organs and better developed olfactory lobes in the brain than do the females of the same species. There is no doubt that their enhanced sense of smell enables the males to find the females in the unlit sea. After they become parasitic their nostrils degenerate.

(From *The Amazing World of Animals* edited by Sir Peter Scott published by Nelson.)

Questions

1. Write down the meanings of the words underlined in the text.

2. Give reasons why marine biologists believed that there was a lifeless zone in the deep sea.

3. Explain briefly why it is an advantage for deep sea fish to have:
a. a swim bladder;
b. huge teeth and enormous jaws;
c. luminous lures;
d. males which become attached to the females;
e. large nasal organs and olfactory lobes.

Summary
1. Most animals and plants are *unevenly distributed* in a habitat.

2. Explanations (*hypotheses*) for their distribution can be tested by using *controlled experiments*.

3. Important factors which decide where animals and plants are found include *dispersal* (could it get there?) and *colonisation* (could it survive and multiply when it arrived?).

4. Successful colonisation depends on the habitat providing enough *food, oxygen and water*.

5. Colonisation may be prevented by *competitors* and *predators*.

Chapter 13 **How living things affect one another**

In a habitat, the plants and animals depend on each other and affect each other: we say that they *interact*.

Animals depend on plants

Green plants are able to make their own food by using sunlight. They are called *producers* because they produce all the materials they need by photosynthesis.

Animals are unable to use the sun's energy, so they have to eat plants, or other animals which may have eaten plants. They are called *consumers* because they consume other living organisms to obtain the materials they need.

Food passes from plants to animals in a chain, called a *food chain*. We can show it like this:

$$\text{plants} \longrightarrow \text{animals}$$
$$\text{(producers)} \qquad \text{(consumers)}$$

The arrow in the food chain means 'eaten by'. For example

$$\text{grass} \longrightarrow \text{cow} \longrightarrow \text{human}$$

shows that cows eat grass and in turn are eaten by humans.

As you can see, human food comes to us by way of food chains. Copy Table 13.1 into your notebook. Write all the food materials you have eaten today in the first column. Next write down beside each food the animal or plant from which the food was made. Then list any animals or plants which were consumed by the animal you ate.

Table 13.1

Food	obtained from:	which ate:
Bread	Wheat	
Milk	Cow	Grass

185

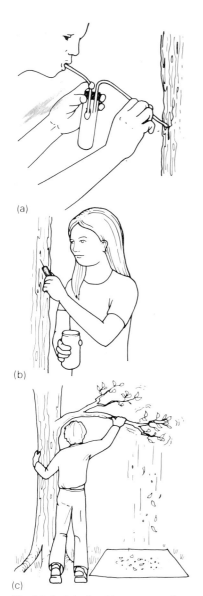

(a)

(b)

(c)

Fig 13.1 (a) Sucking up small animals with a pooter (b) Probing a crack with a pointed instrument (c) Collecting animals from leaves and branches

Experiment 13.1   The inhabitants of a tree

A tree is a major producer; it provides food for a large number of consumers. It may also provide a good place for other producers to live. In this experiment you should choose a large tree and identify the animals and plants living in, on and under it.

You will need a pooter for catching small animals (see Fig 13.1(a)), a pointed instrument for investigating cracks (see Fig 13.1(b)), a sheet or newspaper for collecting animals from leaves and branches (see Fig 13.1(c)), small containers for the animals and polythene bags for the plants.

When you have chosen your tree find out its name and then make five separate collections.

Collection 1. The base of the tree, where it begins to spread out to form the root system. Collect any animals you find, and identify the plants. Mosses often grow in the cracks at the base of trees.

Collection 2. The trunk. Green powdery growth on the bark is a single-celled alga called *Pleurococcus*. If your school is in the country, lichens may grow on the trunk. They consist of two kinds of living organism growing together, an alga and a fungus.

Collection 3. Cracks, forks and hollows. Look carefully but do not damage the bark. The animals living here are often very well camouflaged. If the forks and hollows contain water, you may find some aquatic animals there. Do not forget to include the plants: mosses, ferns and even flowering plants may be found in tree forks.

Collection 4. Branches, twigs and leaves. Examine the twigs and leaves carefully; you may find caterpillars, aphids (greenfly), leaves with marks, spots or tracks on them or galls (large solid lumps on leaves or twigs). Place a sheet under a low growing branch, or spread newspapers beneath it. Shake the branch hard for several minutes and collect any animals which fall on to the sheet.

Collection 5. The dead leaves, twigs and fruits which cover the ground under the tree make up the *litter*. Fill a polythene bag with litter, noting any plants, fungi or small animals which you see.

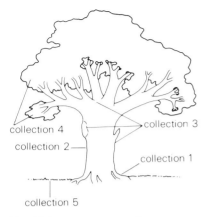

collection 4

collection 2

collection 3

collection 1

collection 5

Fig 13.2 Habitat map of a tree

Before you leave the tree, make a habitat map to show where your collections were made (see Fig 13.2). Back in the laboratory, identify the organisms in your collection using the key in the Appendix. List the plants and animals found in each part of the tree.

Questions for class discussion

1. Did you see signs of any other animals in or on your tree? There may have been holes made by beetles or birds, droppings, pellets or nests of birds or bits of acorns left by squirrels. Add these animals to your list.

2. Which part of the tree had the largest number of consumers feeding on it? Suggest one reason for this observation.

3. Select one of the consumers from this part of the tree and explain how it obtains its food.

4. Select one of the producers living on the tree and explain how it obtains its food.

5. Different trees contain different numbers and different kinds of animals and plants. Why?

Carnivores and herbivores

All animals are consumers, but different animals eat different things. Animals such as ladybirds and spiders are called *carnivores* because they mainly eat the flesh of other animals. Greenfly and woodlice feed on the tree itself and are called *herbivores* (plant eaters). Animals such as humans, which eat both flesh and plants, are called *omnivores*. Figure 13.3 shows eight different animals with their food. List the animals which are carnivores, herbivores and omnivores.

Now return to your list of animals from the tree (Experiment 13.1) and with help from your teacher, write 'carnivore', 'herbivore' or 'omnivore' beside each one.

You can now make three-link food chains for the tree community. Two examples are shown in Fig 13.4.

187

Fig 13.3 Classify these animals as carnivores, herbivores or omnivores

Fig 13.4 Oak tree food chains

Can you write down any more? Remember that the arrow always points to the consumer, showing the direction in which the food passes through the food chain.

Food webs

The oak tree was providing food for several different consumers, so we can combine several food chains into one. (Fig 13.5).

Fig 13.5 Branched food chain

Although ladybirds feed almost entirely on greenfly, spiders will eat a wide range of foods including greenfly and ladybirds as well as woodlice. Where animals feed in more than one food chain, we join the chains to make a *food web* (Fig 13.6).

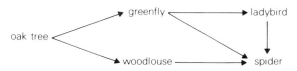

Fig 13.6 Simple food web

The food web can be further extended if we introduce a larger carnivore such as a blue tit. This in turn may be eaten by yet another carnivore such as a hawk. The final carnivore in a food chain is known as the *top carnivore* (Fig 13.7).

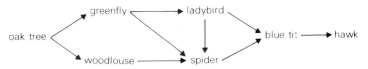

Fig 13.7 Food web with top carnivore

189

Copy this food web into your notebook, adding one more herbivore (for example, a caterpillar), one carnivore (for example, song thrush) and one top carnivore (for example, owl).

Homework assignments

1. Why is a tree called a 'producer'? Write down the names of four more producers which all grow in the same habitat as a tree.

2.
a. Name an animal which feeds as a herbivore and state its normal food.
b. Name an animal which feeds as a carnivore and state its normal food.

3. What is a food chain? Why must it begin with a plant?

4. Write down the names of four producers and four consumers found in your aquarium. Join some of these producers and consumers together in a food chain.

5. In an area of woodland there are foxes, grass and rabbits.
a. Place these organisms in their correct order in a food chain.
b. Add the name of one more herbivore to make a branched food chain.
c. Make a food web by adding the name of a carnivore which feeds on both the herbivores.

Decomposers

When animals and plants die, they rot (decay). You may have wondered what is happening to the dead remains. An important group of living organisms, mainly small animals, fungi and bacteria, feed on the dead remains of other living things. Because they break the material down we call them *decomposers*.

A large deciduous tree may shed hundreds of kilograms of leaves, twigs, bark and other material each year. You collected some of this material (called *litter*) in Experiment

13.1. Now you can find some of the animals which live in it by using a *Tullgren funnel*.

Experiment 13.2   Extracting the animals from leaf litter.

1. Cut out a piece of stiff paper to make a cone. The narrow end of the cone should be about 5 mm wide.

2. Pour a little water into a beaker and fit the narrow end of the cone into the beaker as shown in Fig 13.8.

3. Tip the litter carefully into the cone and switch on the lamp.

4. Leave the apparatus for 24 hours.

5. Carefully tip the contents of the beaker into a petri dish and examine under a lens or binocular microscope.

6. Use the drawings in the Appendix to identify the animals. Write down the names and number of each type of animal caught.

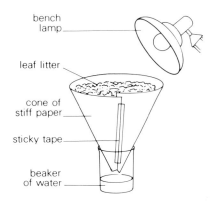

bench lamp

leaf litter

cone of stiff paper

sticky tape

beaker of water

Fig 13.8 A simple Tullgren funnel

Questions for class discussion

1. The heat, light and dryness produced by the light cause the animals to move away and fall into the beaker. Why do small invertebrates mostly move away from these conditions?

2. Which producer provides food for the animals which you extracted with the Tullgren funnel?

3. Some of the animals you caught will be decomposers, but there may have been some carnivores as well. What is their food?

4. At what time of the year would you expect to find the largest number of animals in the leaf litter of deciduous trees? Explain your answer.

5. Table 13.2 shows some of the invertebrates found in 2 kg samples of leaf litter (dead leaves, twigs and soil under the trees) taken from three different habitats.

191

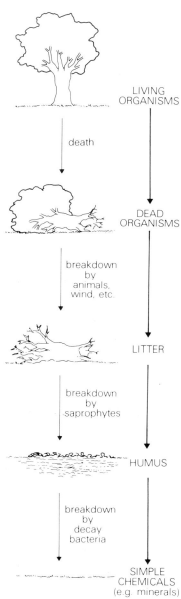

Table 13.2	Woodlice	Mites	Spiders	Snails	Centipedes
Beech wood	27	6	12	14	1
Mixed scrub	68	19	10	57	6
Pine wood	1	5	9	0	1

a. Describe very carefully how you would obtain the leaf litter samples from the habitats.
b. Suggest two reasons why pine wood leaf litter contains fewer invertebrates than the other habitats.

The food cycle

The animals which you extracted from leaf litter by using the Tullgren funnel play an important part in beginning the breakdown of the remains of dead animals and plants (see Fig 13.9).

If you examine rotting wood, fruit or dead animals or the waste products of animals such as 'cow-pats', you can often see signs of another important group of decomposers – moulds or fungi. The surface may be 'mouldy' and covered with fine threads. You may see spore producing structures such as toadstools. These decomposers are called *saprophytes* and they obtain energy and material for growth from the dead remains on which they feed. As a result, the dead remains are turned into a sticky black material in the soil called *humus*. The breakdown of humus is completed by *decay bacteria*. It has been estimated that the number of bacteria in a teaspoonful of soil is greater than the human population of the world.

The end product of the decomposition and decay of living organisms is the release of the chemical materials which they contain so that they can be used again by other living organisms. If decomposers did not exist, the soil would quickly be covered with the dead remains of numerous animals and plants. Growing plants would not be able to obtain the essential chemicals they require, and animals would soon run short of food.

The materials in the living organism are thus passed along a cycle and eventually find their way back into the soil ready to be used again (Fig 13.10).

Fig 13.9 The stages in decay of a tree

Fig 13.10 The food cycle

Questions for class discussion

1. What is humus? Why do farmers spread manure on their fields?

2. Peat consists of the incompletely decayed remains of plants and it is found in wet, acid soil such as that on moorland. Suggest possible reasons why complete decay does not take place in such areas.

Experiment 13.3 Decomposition and decay

Most human food is made from the dead remains of living things. Unless it is preserved in some way, it will decay.

1. Take a thick slice of white bread and damp it with tap water. Do not make it too wet.

2. Place the bread inside a polythene bag and seal it tightly.

3. Label your bag and leave it in a warm place.

4. Examine the bread regularly without opening the bag and make a series of sketches to show the gradual growth of moulds on the surface.

5. After a week or two, take the bread out of the bag and examine it under a lens or a binocular microscope.

6. Remove a very small piece of mould from the bread and place it on a glass slide. Add a drop of water and a cover slip. Examine the slide under the microscope and draw what you see. The mould is a fungus.

Questions for class discussion

1. Plants are green, and make their own food by photosynthesis. Fungi are not green so they have to use ready made food. How is the bread fungus feeding?

2. How does the structure of the fungus help it to obtain food?

3. How many different kinds of fungus did you see on the bread?

4. Suggest two ways by which the fungus might have got into the polythene bag.

5. Describe the changes you would expect to see if the damp bread was left in a sealed container for a long time. What would be found in the container when it was eventually opened after several years? Would the mass of the contents of the bag be the same as the original mass of the piece of bread?

6. In what conditions could you keep bread for a week without it going mouldy? Explain why decay and decomposition is prevented by keeping bread in these conditions.

Homework assignments

1. Explain why decomposers are so important. What would be the result if they all disappeared?

2.
a. Name a saprophyte that you have studied.
b. Describe one way in which it reproduces.
c. On what did you grow it in the laboratory?
d. What makes saprophytes so important in natural communities?

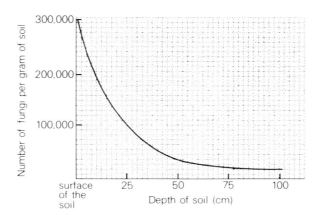

Fig 13.11 Fungi in soil

3.
The graph in Fig 13.11 shows the number of fungi in the soil at different depths from the surface.
a. Where is the greatest number of fungi found?
b. What are the fungi using for food?
c. Why are there fewer fungi further down in the soil?

4.
a. Describe how you could find how many small invertebrates there were in the leaf litter beneath a forest.
b. The graph in Fig 13.12 compares the numbers of mites and other arthropods in leaf litter throughout the year.

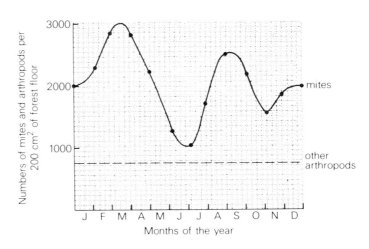

Fig 13.12 Mites and arthropods in leaf litter

195

Suggest one reason why the number of mites varies from month to month while other arthropods remain constant. Briefly describe how you might test your hypothesis.

Maintaining the balance between animals and plants

Consider the simple food chain:

$$\text{grass} \longrightarrow \text{cow} \longrightarrow \text{human}$$

How many of each organism are needed for the chain to survive?

In a lifetime, a person might eat several hundred cows, and each cow would have to eat several million blades of grass. As you go along the food chain, the number of organisms usually decreases and their size usually increases.

If one of the links in the food chain is removed, the whole chain will be affected. The disease myxomatosis (see Chapter 11) which almost eliminated rabbits in some parts of Britain in 1955, affected many other species. Can you suggest what happened to buzzards and hawthorn trees in the following food chain when myxomatosis killed the rabbits?

$$\text{hawthorn seedlings} \longrightarrow \text{rabbits} \longrightarrow \text{buzzards}$$

All the living organisms in a habitat form a community. Look at the stool in Fig 13.13. It will balance only if the legs are the right length. In the same way, the correct numbers of different kinds of animals, plants and decomposers, and the right kind of environment, are required for a *balanced community*.

Fig 13.13 A balanced community

Parasites

Some species of living organisms do not live on their own; they need another species to live with – a sort of partnership. If the partnership is one sided, with only one of the organisms benefiting and the other being harmed by the association, it is called *parasitism*.

Parasites are animals or plants which live inside or on the outside of other organisms. The organism they live on is called the *host* and is the main source of their food. All parasites cause their hosts some sort of damage.

Many human diseases are caused by parasites. Viruses, which cause such diseases as the common cold, and bacteria live inside the human body. They can spread from person to person mainly through the air or by contact.

Experiment 13.4   Looking for parasites

Most living organisms have parasites which live on them. Here are two examples:

1. *Plant galls*. You may have noticed outgrowths on the twigs or leaves of the tree you have studied (Experiment 13.1). Some of these will be galls, caused by a parasitic wasp. Open them carefully with a sharp knife and look for a tiny grub inside. You may find that the grub has left the gall leaving only the tunnel through which it escaped. The life-cycle of a gall wasp is shown in Fig 13.14.

2. *Fish tape-worms*. If you have a pond near your school, the small fish in it may contain tape-worms. Open a freshly dead stickleback or minnow and examine the contents of the body cavity. The segmented white tape-worms inside are sometimes as long as the host. The life-cycle of these tape-worms is shown in Fig 13.15.

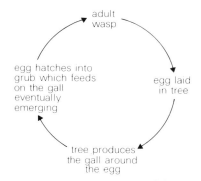

adult wasp

egg laid in tree

tree produces the gall around the egg

egg hatches into grub which feeds on the gall eventually emerging

Fig 13.14 The life-cycle of the gall wasp

Symbiosis

Sometimes two different kinds of organism live together in a partnership from which *both* organisms benefit. This is called *symbiosis*. A lichen is made up of threads of fungus wrapped round cells of an alga. Neither the fungus nor the alga can survive on its own, but together they can live in the most inhospitable climates on earth.

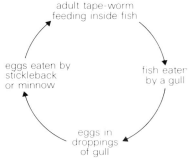

adult tape-worm feeding inside fish

fish eaten by a gull

eggs eaten by stickleback or minnow

eggs in droppings of gull

Fig 13.15 The life-cycle of the fish tape worm

Questions for class discussion

1. What special problems does a parasite such as a tape-worm have, which other organisms do not? How are they overcome?

2. Make a list of some important diseases, and the parasites which cause them.

3. It is sometimes very difficult to discover whether two organisms are living together as parasite and host or to the benefit of both organisms. Many kinds of bacteria live inside the human intestine making use of our food. Some may provide us with useful chemicals. How could a scientist show whether these bacteria are useful to us or parasitic upon us?

Homework assignments

1. Match the descriptive terms in the first column with the most appropriate example in the second column. Write down the pairs in your notebook.

Saprophyte	Oak tree
Parasite	Toadstool
Dead plants	Sheep
Carnivore	Litter
Herbivore	Tape-worm
Producer	Fox

2.

a. Why do most communities contain fewer carnivores than herbivores?

b. What would happen to the number of

(i) carnivores

(ii) producers

if all the herbivores in a community were shot?

3.

a. Write down the name of a habitat that you have studied.

b. From this habitat write down the name of one example each of:

a decay organism (saprophyte)	a herbivore
an arthropod	a carnivore
an animal without a hard skeleton	a green plant

4.
a. Name the *three* most common kinds of plants found in a habitat you have studied.
b. Give a clear description of this habitat.
c. List *three* different ways in which the plants you mention are of benefit to animals in the habitat.
d. If most of the plants had suddenly died out, say what you think would happen to the remaining animals, and why. Give examples.
e. Give *two* possible reasons why the plants in such a habitat might suddenly start to die out.

5. What are the main differences between a saprophyte and a parasite? Name one example of each and describe how they feed.

6. On a particular coral reef, sea slugs feed on sea anemones; sea anemones catch rock-fish; copepods (small shrimp-like creatures) feed on plankton (small floating plants) and the copepods are eaten by rock-fish.
a. Which are the producers?
b. Which are the top carnivores?
c. The plankton feed like normal plants. How do they get their food?
d. Write down the food chain for this coral reef.

7. Pondweed   caddis   stickleback   pike
a. The above are all part of a food chain. Draw the chain in your notebook.
b. Which of the organisms named above is   (i) a herbivore,   (ii) a top carnivore,   (iii) a producer?
c. If all the animals in the food chain were collected and their total masses found, which level would have the most mass: caddis, stickleback or pike? Explain your answer.

8. Write out the following sentences using the phrase which completes them most accurately.
a. Humus is
    found in fertile soil
    essential for photosynthesis
    a bone in the arm
    destroyed by light

b. When left uncovered, food will decay because it
        contains parasitic bacteria
        reacts with the air
        contains parasitic viruses
        is consumed by bacteria
c. A parasite is an organism which
        feeds only on dead materials
        only attacks plants
        kills its prey and then feeds on it
        feeds on living material
d. Lichens are
        dead plants
        saprophytes (decay organisms)
        an association between an alga and a fungus
        plants which live as parasites on other plants

9.
a. Describe the main features of a habitat you have studied.
b. Of all the organisms found there, which do you consider to be the most interesting plant or animal? Give reasons for your choice.
c. Draw a simple food web for this habitat, including not less than *eight* clearly named organisms.

10. In an experiment to investigate the food of the common earthworm, *Lumbricus terrestris*, 100 discs of the same size were cut from oak leaves, and another similar set from beech leaves. The discs were scattered at random on a grass plot which was surrounded by a wooden frame and covered by a lid of nylon mesh. Each month, from July to January, the frame was examined and some discs were found to have been eaten. The number of oak leaf discs and of beech leaf discs that remained were counted and the number of each was made up once more to 100. The results in Table 13.3 were obtained:

a. Draw a graph to compare the consumption of beech and oak discs.
b. Why was the number of leaf discs made up each month to 100?
c. Suggest why more leaf discs of both kinds were eaten in September and October than in any other months.

Table 13.3

Month	Number of beech discs eaten	Number of oak discs eaten
July	33	8
August	5	2
September	8	30
October	10	25
November	5	15
December	0	0
January	1	3

d. Suggest why so few discs were eaten in December and January.

Background reading

Ants are said to outnumber all other land-dwelling animals. There are more than 6000 species found from the arctic to the equator, from deserts to seashores. Large colonies may contain as many as half a million ants of three different types. Some are males, some females (queens) and the majority are infertile females (workers). Only the male and queen ants have wings.

At special times of the year the new generation of males and queens leave the nest and mate in mid-air in what is called the 'marriage flight'. After this the males die and the queens lose their wings and go off to found new colonies.

In the garden ant (*Lasius niger*) the fertilised queen burrows into the ground and starts to lay eggs. Over her lifetime of 6–7 years she may lay thousands of eggs. She feeds the larvae when they hatch on her own sugary saliva. The first larvae metamorphose into workers and start to feed the queen. They also protect the nest and turn and clean the eggs. The larvae are fed by the workers with partly digested food from their crops.

In mid-summer the queen starts to lay fertilised eggs by releasing some of the stored sperm from her marriage flight. When they hatch, these larvae are fed on a specially rich diet by the workers.

Ants may be herbivorous or carnivorous or sometimes both. The three most common species in Britain are the little black garden ant found under stones, the yellow meadow ant found in lawns and fields and the wood ant which makes hills several feet high and is mainly a scavenger.

If you examine an ants' nest carefully you may see 'foreign' species. Ants go out on raids to capture eggs, larvae and pupae from other nests and bring them back to their own. As soon as they have hatched, the slaves seem willing to work for their new masters. Other welcome guests in the ant colony are a white bristletail and a white woodlouse. Both are blind and do the job of 'dustmen' in the colony.

(From *Ants* by Anthony Wootton published by Priority Press.)

Questions
1. Ants may be herbivorous, carnivorous or scavengers. Explain briefly the differences between these methods of feeding.

2. Three phrases in the text have been underlined. Rewrite each phrase, using words which a non-biologist would understand.

3. Describe in your own words the ways in which
a. ants co-operate with each other
b. ants co-operate with other species.

4. Consult books about ants (there are some suggestions in the Appendix) to find out as much as you can about one of the following:
a. Building a formicarium (artificial ants' nest);
b. The 'milking' of aphids by ants;
c. Ants which grow fungus 'gardens';
d. Army ants.

Summary
1. In a habitat, the plants and the animals *interact*.

2. The animals (*consumers*) in a community depend on the plants (*producers*) to provide them with their ready made food.

3. Animals can be divided into meat eaters (*carnivores* or *predators*) and plant eaters (*herbivores*).

4. *Food chains* and *food webs* show the path of food from one organism to another in a community.

5. The dead remains of living organisms and their waste products are broken down by *decomposers* which include animals, fungi (*saprophytes*), and decay bacteria.

6. Some living organisms live in partnerships. In *symbiosis* both partners benefit; in *parasitism* the *host* is always harmed by the *parasite*.

Chapter 14 **The human animal**

As human beings we belong to the Animal Kingdom. Our scientific name is *Homo sapiens*, but we are usually referred to simply as humans or Man.

What group of animals do we belong to?

We belong to the mammals, which is one of the groups of backboned animals or vertebrates (see p. 36). A human is shown with some of his mammalian relatives in Fig 14.1.

Mammals possess two main features (see p. 1C2): they have *hair*, and the young are fed on *milk* which is produced by the mother's *mammary glands*. This applies as much to humans as to other mammals.

Fig 14.1 The human with some mammalian relatives

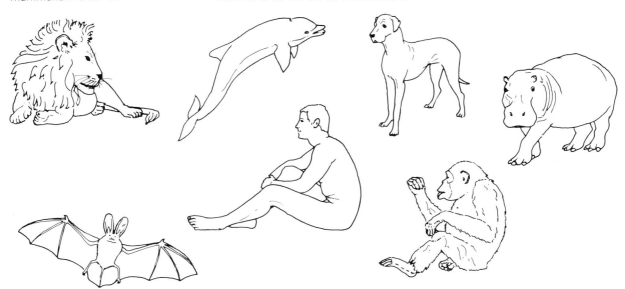

Man as a mammal

Young mammals obtain milk by sucking the mother's nipples. Most mammals have many nipples. However, the human has only two.

The main function of a mammal's hair is to keep the body warm. Mammals are warm-blooded and it is important to prevent heat being lost. Humans have less hair than most other mammals, but they make up for this by wearing clothes.

We have many other features in common with our mammalian relatives, but we also differ from them in a number of ways. For example, we walk on only one pair of legs, that is we are *bipedal*. Most mammals walk on all-fours, that is they are *quadrapedal*. Being bipedal means that we can use our arms and hands for other jobs, such as making things and writing.

Another way in which we differ from other mammals is in what our brain can do. All mammals have complicated brains but humans have the most advanced brain of all, and we use it to reason things out and solve problems. Moreover, we can communicate with each other by means of language. It is largely because of our brain, and our ability to communicate, that we can live in an advanced technological society.

One more way in which the human differs from other mammals is that the young are born at a stage when they are completely helpless, and it takes them a relatively long time to develop to maturity. During this time they learn from their parents and others how to fend for themselves. No other mammal goes through such an elaborate period of training and takes such a long time to reach adulthood.

Questions for class discussion

1. Make a list of the ways in which you are similar to, and different from, a mammal such as a dog. Why are humans and dogs both included in the mammal group?

2. Why do you think humans have fewer nipples than most other mammals?

3. What are the advantages of being bipedal? Can you think of any other mammals which are at least partly bipedal?

4. Look at Fig 14.1. Which of the mammals in this picture is most like the human? Give reasons for your choice. Why are they all regarded as mammals?

5. Someone has said that the human brain not only solves problems but also *creates* problems. What world wide problems have been created by the human brain?

The parts of the body

The human body consists of three main parts: the *head*, *trunk* and *limbs* (Fig 14.2). The head is joined to the trunk by the *neck*. The trunk is divided into two parts, the *thorax* and the *abdomen*. The limbs are the arms and legs.

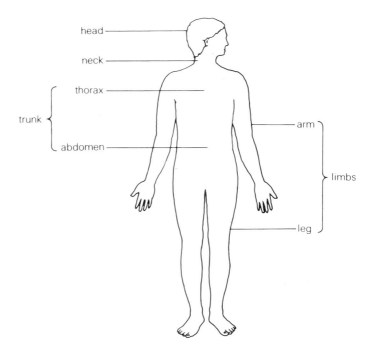

Fig 14.2 The main parts of the human body

Organs

Inside the body there are all sorts of important structures. We call these *organs*. You have already met some of them in previous chapters. Each organ has one or more special jobs to do.

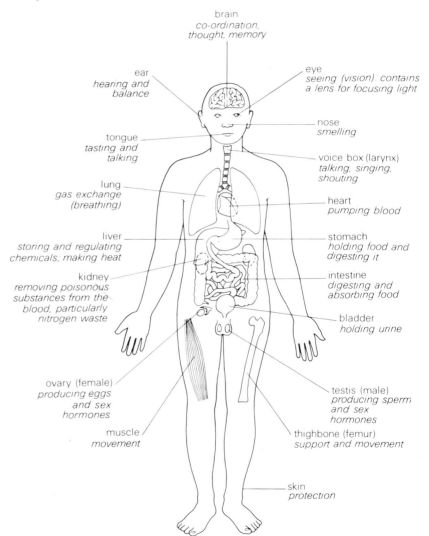

brain
*co-ordination,
thought, memory*

ear
*hearing and
balance*

eye
*seeing (vision): contains
a lens for focusing light*

nose
smelling

tongue
*tasting and
talking*

voice box (larynx)
*talking, singing,
shouting*

lung
*gas exchange
(breathing)*

heart
pumping blood

liver
*storing and regulating
chemicals; making heat*

stomach
*holding food and
digesting it*

kidney
*removing poisonous
substances from the
blood, particularly
nitrogen waste*

intestine
*digesting and
absorbing food*

bladder
holding urine

ovary (female)
*producing eggs
and sex
hormones*

testis (male)
*producing sperm
and sex
hormones*

muscle
movement

thighbone (femur)
support and movement

skin
protection

Fig 14.3 Some of the more important organs of the human body, and their main functions

The main organs found in the human body are shown in Fig 14.3. With each label there is a brief note on what the organ does.

Questions for class discussion
These questions relate to Fig 14.3.

1. Certain organs are at, or close to, the surface of the body. Which ones are in such a position, and why?

2. One of the organs can be said to 'store information'. Which one?

3. Where does blood go after it has left the heart?

4. What gases are exchanged in the lungs?

5. Which organs are joined to one another, and why?

6. Which organs occur in pairs? Why is this useful?

7. An important function of the skin is 'protection'. What does the skin protect us from?

8. Can you think of any other organs which are not shown in Figure 14.3?

9. One of the organs contains a lens. Which one?

10. Which organs occur in only one place, and which ones are dispersed throughout the body?

Homework assignments
1. Write down one feature of the human which is:
a. common to all living things;
b. common to all animals;
c. found in all vertebrates but not in any other group;
d. found in all mammals but not in any other group;
e. different from any other mammal.

2. Each of the following statements applies to at least two different organs in the human body. Write down the names of all the organs to which each statement applies.
a. They get rid of unwanted substances from the body.

b. They store substances.

c. They enable us to walk around.

d. They produce gametes.

e. They tell us about our environment.

3. The pictures in Fig 14.4 are of five organs found in the human body. Give the name of each one, say whereabouts it occurs, and write down one function which it carries out.

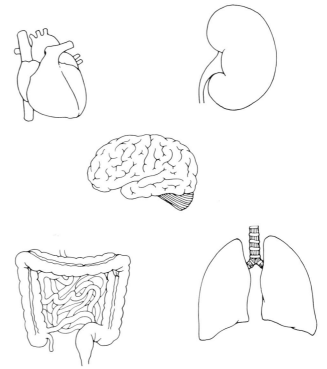

Fig 14.4

4. In Fig 14.5, the human body is divided into eight parts labelled A to H.

a. Which parts contain the thorax, and which ones contain the abdomen?

b. Name one organ which is found only in each of the parts labelled A to E.

c. In which part or parts would you find   (i) the larynx   (ii) the muscles   (iii) the kidneys?

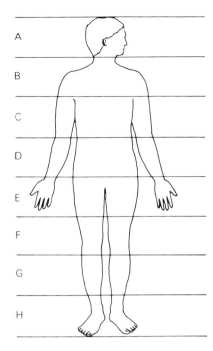

A

B

C

D

E

F

G

H

Fig 14.5

5. Some of our organs work together and their jobs are closely interlinked. Arrange the organs shown in Fig 14.3 into pairs according to how close they are in the jobs they do. In each case explain why you have linked them together.

6. The human is generally regarded as the most advanced of all mammals. Write down those features of the human species which, in your opinion, have made it so successful.

Background reading

Imagine a piece of land twenty miles long and twenty miles wide. Picture it wild, inhabited by animals small and large. Now visualise a compact group of sixty human beings camping in the middle of this territory. Try to see yourself sitting there, as a member of this tiny tribe, with the landscape spreading out around you farther than you can see. No one apart from your tribe uses this vast space. It is your exclusive home range, your tribal hunting ground. Every so often the men in your group set off in pursuit of prey. The women gather fruits and berries. The children play noisily round the camp site, imitating the hunting techniques of their fathers. If the tribe is successful and swells in size, a splinter group will set off to colonise a new territory. Little by little the species will spread.

Imagine a piece of land twenty miles long and twenty miles wide. Picture it civilised, inhabited by machines and buildings. Now visualise a compact group of six million human beings camping in the middle of this territory. See yourself sitting there, with the complexity of the huge city spreading out all around you, farther than you can see.

Now compare these two pictures. In the second scene there are a hundred thousand individuals for every one in the first scene. The space has remained the same. It has taken a mere few thousand years to convert scene one into scene two. The human animal appears to have adapted brilliantly to his extraordinary new condition. This civilising process has been accomplished entirely by learning and conditioning. Biologically he is still the simple tribal animal depicted in scene one.

(Adapted from *The Human Zoo* by Desmond Morris, published by Jonathan Cape.)

Questions
1. What do you think the author means by 'learning and conditioning' in the second to last sentence?

2. Make a list of as many ways as you can think of in which humans show that biologically they are still simple tribal animals.

3. What major advances have made it possible for a hundred thousand individuals to inhabit an area previously occupied by one individual?

Summary

1. Humans belong to the group of animals known as *mammals*.

2. In common with other mammals, humans are warm-blooded and have *hair* on their bodies. They have *mammary glands* with which to feed the young.

3. Humans differ from other mammals in being *bipedal*, in having the most advanced brain, and in giving birth to young that are helpless at first and take a long time to reach maturity.

4. The human body consists of three main parts: *head*, *trunk* and *limbs*.

5. The human body contains many important structures known as *organs*.

Chapter 15 **Nutrition in the human**

How many meals do you have each day, and how much food do you eat? Estimate the approximate mass of food in grams which you eat in one day. Then calculate the mass of food which you would consume in, say, twenty years.

Diet

What sort of food do you eat at each meal? If you were to make a list it would probably include things such as corn-flakes, eggs, bread, meat and so on. All these make up our *diet*.

However, there is a more scientific way of describing our diet, and that is to list the different chemical substances that our food contains. They are these:

carbohydrates   vitamins
fats   salts
proteins   water

It is most important that our diet should include all these chemical substances in sufficient amounts – and the same applies to other animals too. They are essential for good health and if one or more of them is lacking from our diet, we may become very ill. One of the problems in poorer countries is that many people do not get as much of these substances as they need.

Now let us see what these substances are needed for.

Carbohydrates

Carbohydrates give us energy. The two main carbohydrates are *sugar* and *starch*. Figure 15.1 shows a few examples of foods that contain sugar and starch.

Fig 15.1 Four foods which contain relatively large amounts of carbohydrate.

Sugar dissolves in water; in other words it is *soluble*. In contrast, starch will not dissolve in water: it is *insoluble*.

You can usually tell if a particular food contains sugar because it tastes sweet. To find out if starch is present, a simple chemical test can be carried out. You used this test on an earlier occasion to find out if a green leaf contained starch. Can you remember what the test was? (See p. 108).

Experiment 15.1   To find out if starch is present in various foods

Try this test on a small piece of bread.

1. With a pipette place one or two drops of dilute iodine solution on the bread. If you get a blue-black colour, starch is present.

2. Now try this test on other foods which your teacher will give you. Make a list of the foods in your notebook, and in each case say whether or not starch was present.

Questions for class discussion

1. Foods containing starch come mainly from plants, whereas foods without starch come mainly from animals. Is this true of the foods you tested?

2. Why is it useful to know if a particular food contains starch?

Fats

Fig 15.2 Some fatty foods

We are all familiar with fatty foods: some examples are shown in Fig 15.2. Fat is important for two main reasons:

1. It gives us energy, particularly when the body runs short of carbohydrate. Fat is stored in the body mainly under the skin.

2. The fat under the skin serves as an insulator, cutting down heat loss in cold weather. It thus helps to keep us warm.

Animals such as whales and seals, which live in very cold water, have a particularly thick layer of fat: this is known as blubber.

Experiment 15.2   To find out if there is any fat in a food

Try this with some butter or lard.

1. Rub the food on a piece of brown paper. If this leaves a transparent mark on the paper even when the paper dries, it means that fat is present.

2. Now try this test on other foods given to you by your teacher. Record the results in your notebook.

Questions for class discussion

1. Fat is sometimes described as an 'energy store'. Why is it useful for an animal to have an energy store in its body?

2. Two boys dive into an unheated swimming pool. One of the boys is fat, the other thin. Which one is likely to lose heat quicker? Give reasons for your answer.

Proteins

Figure 15.3 shows some foods which are particularly rich in protein. We need proteins for two main reasons:

1. It forms the main structures in the body. For example, our muscles, skeleton and cells all contain protein. Protein is the basic material in every living cell.

2. It helps to regulate the chemical reactions that go on inside the body. The proteins that do this are called *enzymes*.

Proteins also provide energy, particularly if carbohydrate and fat are in short supply.

Questions for class discussion

1. It is particularly important that pregnant mothers and growing children should get plenty of protein. Why?

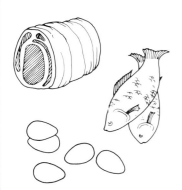
Fig 15.3 Some protein containing foods

2. Adults need protein to replace worn out or damaged cells. In fact a process of continual replacement of materials goes on in the body all the time.
a. Think of as many ways as you can in which cells may get damaged.
b. Why do you think cells wear out eventually?

Vitamins

Vitamins are special substances which are essential for good health. Most of them are needed in very small amounts. Each one performs a particular function, and if any of them are missing from the diet, illness and even death may result.

The different vitamins are known by letters (A, B, C etc.). As an example let us take vitamin C. This vitamin helps our cells to stick together. If it is missing from the diet, the person gets a disease called scurvy. One of the signs of this disease is that the gums bleed around the teeth.

Vitamin C is abundant in citrus fruits such as oranges and lemons. It is also found in green vegetables. It is easily destroyed by prolonged heating.

In general fresh fruit and vegetables are a good source of vitamins. One of the best animal foods for vitamins is liver.

Questions for class discussion
1. In Admiral Nelson's day, limes were included in ships' rations. Why do you think this was done?

2. A hundred years ago diseases caused by lack of vitamins were common in Britain. Nowadays such diseases are rare. What has brought about this change?

3. It is unwise to overcook cabbage. Why?

Salts

For healthy life the body requires the salts of certain mineral elements such as calcium, phosphorus and iron. For example, calcium and phosphorus are needed to make bones hard. Iron helps our blood to carry oxygen round the body; this is because it is needed for the formation of haemoglobin, the red substance in the blood.

Like vitamins, mineral salts are needed in only small amounts, but if any are missing from the diet the person may become very ill. Figure 15.4 shows what happens if a baby does not get enough calcium: the bones are soft and may bend. Like vitamins, salts are particularly plentiful in fresh fruit, vegetables and liver.

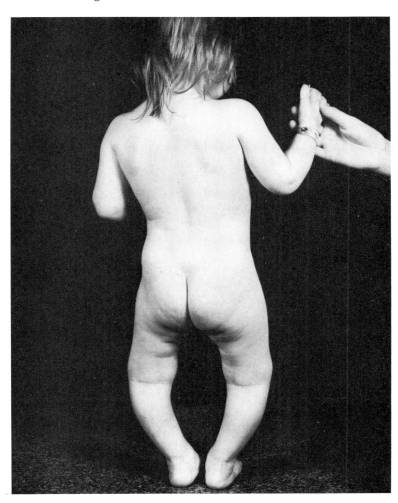

Fig 15.4 The bones of this child are bent as a result of a lack of calcium in the diet. This disease is called rickets

Fig 15.5 At the top is a pot plant which was not watered by its absent-minded owner. Below is the same plant shortly after watering

Questions for class discussion

1. The disease shown in Fig 15.4 is called rickets. It is caused by lack of calcium in the diet. The same disease can be caused by lack of a certain vitamin. Find out which vitamin is needed for hardening bones, and what sort of foods contain it.

2. Sometimes people who are tired and lack energy take iron tablets. In what way might this help them?

Water

Water is an extremely important part of all living things. In fact the human body is about 70 per cent water. There are many reasons why water is so important. Here are three of them:

1. Many substances which occur in the body are dissolved in water. In solution these substances can come into contact with each other and react together.

2. Being a liquid, water can move easily around the body, carrying things with it. This is important in transporting materials in our blood system.

3. It gives shape and form to the bodies of animals and plants. For example, if a plant loses a lot of water, it may droop. We call this *wilting* (Fig 15.5).

Question for class discussion
You have just read three reasons why water is important to living things. Make a list of as many other reasons as you can think of why water is important. Do not restrict your list to humans; include other organisms as well.

Experiment 15.3   To find out how much water a plant contains
You will need a lettuce which has received plenty of water beforehand.

1. Weigh the lettuce and write down its mass in grams

2. Put the lettuce on a metal dish and place it in an oven at 100°C for an hour.

3. Take the lettuce out of the oven and let it cool down. Then put it in a desiccator overnight.

4. Re-weigh the lettuce. Calculate the percentage loss in the mass.

Questions for class discussion

1. What percentage of the lettuce is water? What assumptions have you made in arriving at your answer?

2. How could you estimate the amount of water in the plant more accurately?

A complete diet

Every day our bodies use a certain amount of carbohydrate and other food substances. How much of each substance the body uses will depend on many things. For example, an energetic person will use up a lot of carbohydrate, and a growing child will use up a lot of protein.

The amount of each substance used must be replaced in the person's food. A diet which achieves this is called a complete diet. A complete diet provides all the substances the body needs.

People that organise the feeding arrangements in schools and hospitals must make sure that a complete diet is provided. To do this they need to know how much of each essential substance occurs in different foods, and how much of each substance we need per day. This information can be found in books on nutrition.

Homework assignments

1. 'An apple a day keeps the doctor away'. Explain the reasoning behind this well known saying.

2. You are a scientist working in a remote village in a developing country. The inhabitants are suffering from an unknown disease which you *think* might be caused by the lack of a certain vitamin in their diet.

a. What would you do to test your idea?

b. Assuming that your idea turns out to be right, what remedy would you recommend?

3. A scientist fed some mice on a diet consisting of water, salts, carbohydrates, fat and protein. After a while their hair became sparse.

a. What might have caused their hair to become sparse?

b. Describe an experiment which could be done to find out if your answer is correct.

4. Write a letter to the newspaper either supporting or condemning the use of animals in experiments of the kind outlined in question 3. (In Britain it is illegal to perform experiments on animals such as mice without a special licence.)

What happens to the food we eat?

Much of our food is in an insoluble form. For the body to use it, the insoluble substances must be turned into soluble ones. This process is known as *digestion* and it occurs in the *gut*. The gut is really a long tube which extends from the *mouth* to the *anus* (Fig 15.6).

Digestion starts as soon as food is put in the mouth. Here starch is turned into a soluble substance. This conversion is brought about by *saliva* ('spit').

Experiment 15.4   To find out what happens to starch in the mouth

1. Cut a cube of bread about 1 centimetre square and put it in your mouth.

2. Chew the bread without swallowing it.

3. After two minutes spit out a small amount of the food into a beaker, and add a drop of dilute iodine solution. Is a blue-black colour given? If it is, starch must be present.

Fig 15.6 The main parts of the human gut

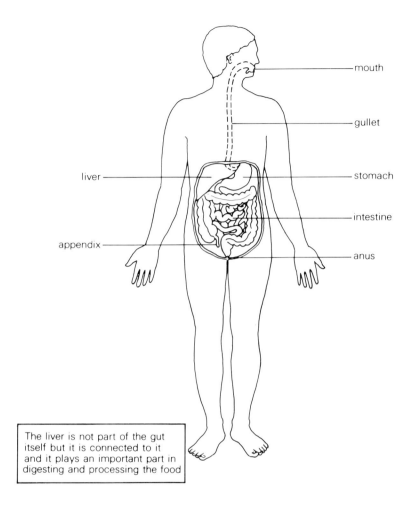

mouth

gullet

liver

stomach

intestine

appendix

anus

The liver is not part of the gut itself but it is connected to it and it plays an important part in digesting and processing the food

4. Continue to chew, without swallowing. After a further two minutes test another sample of the food with iodine. Is any starch still present?

5. Test further samples at two-minute intervals until a blue-black colour is *not* given. When this stage is reached, we conclude that starch is no longer present in the food.

219

1. You may have noticed that towards the end of the experiment the food tasted sweet. What is the sweet taste due to?

2. Write a simple word equation for the chemical change which has taken place in your mouth.

3. Chewing helps to dissolve a solid piece of food. In what way does chewing help?

Enzymes

What is it about saliva which enables it to digest starch? The answer is that it contains an *enzyme*. The enzyme in saliva is able to turn insoluble starch into soluble sugar.

Although digestion starts in the mouth, most of our food is digested in the stomach and intestine which contain many different enzymes. These turn the food into a variety of soluble substances.

Absorption

In the intestine, the soluble substances are absorbed into the bloodstream. They are then transported to all the cells of the body where they are used for producing energy and carrying out other vital functions.

Some substances cannot be digested. They remain in a solid form and pass along the intestine to the anus.

The main processes which occur in the gut are summed up in Fig 15.7.

Questions for class discussion
1. The wall of the intestine contains lots of blood vessels. Suggest two functions which these vessels may perform.

2. The flow of blood to the intestine increases after a meal. Why do you think this is?

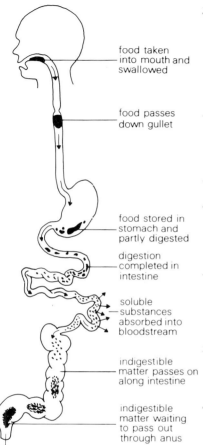

food taken into mouth and swallowed

food passes down gullet

food stored in stomach and partly digested

digestion completed in intestine

soluble substances absorbed into bloodstream

indigestible matter passes on along intestine

indigestible matter waiting to pass out through anus

Fig 15.7 This diagram summarises what happens to our food after it has entered the mouth

3. In humans, the cellulose contained in plant foods such as potatoes and cabbages cannot be digested in the gut. Why can it not be digested? What happens to it eventually? Why is it important?

Homework assignments

1. Starch has to be digested before the body can make use of it.
a. What does the word 'digested' mean?
b. Why does the starch have to be digested?
c. What use does the body make of the starch eventually?

2. Two athletes, A and B, wanted to give themselves some extra energy just before a race. A had a drink of glucose (sugar), and B ate a slice of bread. Which one was the more sensible, and why?

3. When you have not eaten anything for several hours you feel hungry. Put forward two theories to explain how the feeling of hunger is brought on.

Summary

1. A complete diet is made up of carbohydrates (sugar and starch), fats, proteins, vitamins, salts and water.

2. Carbohydrates and fats are energy foods, proteins are for body building, vitamins and salts are for general health, and water performs a variety of functions.

3. Most of the food we eat is insoluble. As it passes through the mouth, stomach and intestine it is *digested*.

4. The soluble substances thus formed are *absorbed* into the blood and taken to the cells.

Chapter 16 **Energy and respiration**

All living organisms need energy in order to continue their activities. Where does this energy come from?

In chemistry you will have learned that when a carbon compound such as candle wax or coal is burned, three things happen:

1. oxygen is used up;

2. water and carbon dioxide gas are formed;

3. energy in the form of heat is released.

The same kind of thing happens inside our bodies. Food substances which we have eaten are *oxidised* and as a result energy is released, water is formed, and carbon dioxide gas is given off:

$$\underbrace{\text{FOOD + OXYGEN}}_{\text{raw materials}} \longrightarrow \underbrace{\text{WATER + CARBON DIOXIDE}}_{\text{products}} \text{ + ENERGY}$$

We call this process *respiration*. Although the raw materials and products are the same as when carbon compounds are burned, respiration takes place much more gently and in a series of small steps. When sugar is burned in a test-tube there is a fizz and a crackle and the temperature shoots up. If this happened in our bodies we would die immediately!

Respiration is important because it gives us energy for running around and playing games. We also need it for growing and for repairing parts of the body which get worn out or damaged, and we need a certain amount of energy just for staying alive. Some of the energy, in fact most of it, is set free as heat, and this helps to keep us warm.

Producing carbon dioxide

The main type of food from which we get energy is sugar, which is a carbohydrate. Carbohydrates contain carbon, hydrogen and oxygen. The carbon dioxide which is given off in respiration comes from the breakdown of the carbohydrate.

Experiment 16.1 To find out if we breathe out carbon dioxide

1. Pour some limewater into a test-tube to a depth of about two centimetres. Limewater is a solution of calcium hydroxide and it turns milky in the presence of carbon dioxide.

2. Breathe out into the limewater, using a glass tube as shown in Fig 16.1. Does the limewater turn milky? What do the results tell you about the air we breathe out?

Questions for class discussion

1. The limewater *might* have been turned milky not by carbon dioxide in the air you breathed out, but by carbon dioxide present in the air around you.

    Describe an experiment which you could do to rule out this possibility. A number of different methods could be used: do not be afraid to put forward your ideas.

2. Why does carbon dioxide make limewater go milky? If you do not know the answer, try looking it up in a chemistry book.

    We have seen that human beings give out carbon dioxide. We will now do an experiment to find out if other organisms do the same.

Experiment 16.2 To find out if germinating seeds give out carbon dioxide

You will need some seeds which are in the process of germinating. French beans will do.

limewater

Fig 16.1 Breathing out through limewater

1. Put some moist cotton wool at the bottom of a flask, and place about twelve germinating seeds on top of the cotton wool.

2. Set up the flask as shown in Fig 16.2. Smear Vaseline over all the joints to make the apparatus airtight.

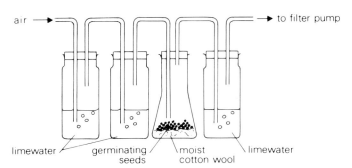

air →   → to filter pump

Fig 16.2 Apparatus for finding out if germinating seeds give out carbon dioxide

limewater   germinating   moist   limewater
             seeds   cotton wool

3. Leave the apparatus for about ten minutes, then start the suction pump and gently draw air through. What happens to the limewater in the third bottle? Does it turn milky? If so, why?

Questions for class discussion

1. What is the point of having the first two bottles of limewater in this experiment? Are they really necessary?

2. Why did you have to leave the apparatus for ten minutes before starting up the suction pump?

3. It is possible that the limewater in the third bottle might have turned milky because the cotton wool in the flask gave out carbon dioxide. Using the same apparatus, how could you rule out this possibility?

4. What experiment could you do to find out if the seeds have to be *alive* to give out carbon dioxide?

5. The results of this experiment suggest that plants, as well as animals, respire. What process in a germinating seed requires energy?

   The apparatus in Fig 16.2 can be used with other small organisms, for example woodlice, earthworms or snails.

You can even use it with a mouse or gerbil, but for such animals you need to have a large jar instead of the flask, and you must start drawing air through the apparatus as soon as you have put the animals in the jar. Why do you think this is necessary?

Homework assignments

1. Living things need energy in order to live.
a. What is the name given to the process which releases energy from food?
b. Write down three ways in which this energy is used by living things.
c. What gas is (i) used up and (ii) given out in this process?

2. Which organs in your body do you think might need a particularly good supply of oxygen when you:
a. are running a race;
b. are taking an examination;
c. are eating a sandwich;
d. have just had your lunch?

3. In an attempt to find out about respiration, a schoolgirl set up the experiment shown in Fig 16.3. She found that at the start of the experiment the water level in the glass tube was at point A. Half an hour later it had risen to point B.
a. Why was limewater placed in the test-tube with the woodlice?
b. Why do you think the water rose in the glass tube?
c. What further experiment should the girl do to be certain that the rise in the level of the water was caused by the woodlice?
d. Another girl set up exactly the same experiment, but found that the water did not rise in the glass tube at all. Suggest reasons for this.

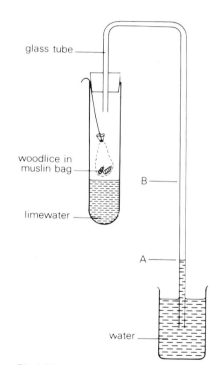

glass tube

woodlice in muslin bag

limewater

B

A

water

Fig 16.3

Breathing

We have seen that we need oxygen for getting energy from food. Carbon dioxide, on the other hand, is poisonous, and

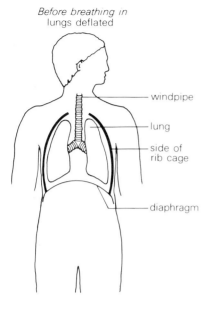

Before breathing in
lungs deflated

- windpipe
- lung
- side of rib cage
- diaphragm

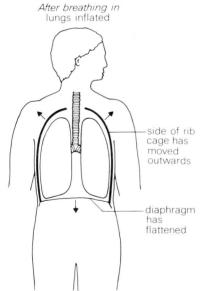

After breathing in
lungs inflated

- side of rib cage has moved outwards
- diaphragm has flattened

Fig 16.4 The chest and lungs before and after breathing in

we must get rid of it as quickly as possible. Both these functions – taking in oxygen and getting rid of carbon dioxide – are achieved by *breathing*.

When we breathe in, air is drawn into our *lungs*. These are complicated sacks situated in the chest. When breathing in, the chest increases in size and air is drawn in. When breathing out, the chest decreases in size forcing air out.

How does the chest increase and decrease in size? The sides of the chest are formed by the *ribs*, and the bottom is spanned by the *diaphragm*, a dome shaped sheet of muscle. When you breathe in, your ribs move outwards and your diaphragm flattens. As a result the chest gets larger (Fig 16.4). When you breathe out, the opposite happens: your ribs move inwards and your diaphragm bows upwards. The result is that the chest gets smaller.

Experiment 16.3   To find out how much air you can take into your lungs

1. Fill a basin with water. Lay a bell jar on its side in the basin, so as to fill it with water. Then place the bell jar in an upright position on supports. Insert a bent tube under the rim as shown in Fig 16.5(a).

2. Take as deep a breath as you can. Then put the end of the bent tube in your mouth and hold your nose. Now breathe out as much as you possibly can: this will displace some of the water from the bell jar as shown in Fig 16.5(b). As soon as you have finished breathing out, take the tube out of your mouth.

3. From the scale on the side of the bell jar, read off the volume of air in litres which you have breathed out.

4. Try this experiment on different people in your class (and on your teacher) and compare the results.

Questions for class discussion

1. There is considerable variation in people's lung capacities. Can you suggest possible reasons for this? Do you think a person's lung capacity is related to his or her size?

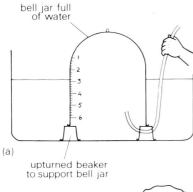

(a)

bell jar full of water

upturned beaker to support bell jar

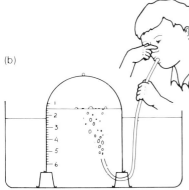

(b)

Fig 16.5 Apparatus for finding how much air you can take into your lungs

Fig 16.6 The main structures that make up the circulation. The arrows indicate the direction in which blood flows

2. In this experiment you have found out the *maximum* volume of air which you can take into your lungs. However, in normal breathing, you take in only a fraction of this amount.

Using the same apparatus, how could you find out what this 'normal' amount is? In what circumstances do we breathe more deeply and thereby take more air into our lungs?

Blood and the circulation

Think of the air we breathe in. Oxygen is taken out of this air and carried to all parts of our bodies. This job is performed by our *blood*.

Our blood is kept moving by the *heart* which pumps it through a series of tube-like *blood vessels* (Fig 16.6).

Vessels which carry blood from the heart to the various organs of the body are called *arteries*. The arteries split up into narrow *capillaries* which then join up again into *veins*. The veins take the blood back to the heart. The heart and the veins contain *valves* which keep the blood moving in one direction and stop it flowing backwards, just as the valves in a foot pump prevent air going backwards.

So the blood flows round and round the body. We call this our *circulation*.

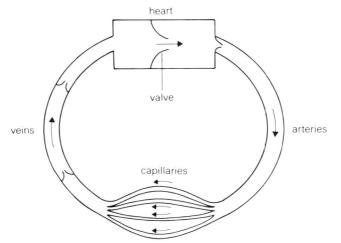

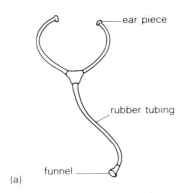

(b)

Fig 16.7 Using a stethoscope for listening to your heart

Experiment 16.4 Listening to the heart

You will need a *stethoscope*: the kind that doctors use for listening to the heart (Fig 16.7(a)).

1. Put the ear pieces in your ears and the opening of the funnel on your chest (Fig 16.7(b)). Listen! Can you hear regular thud-like sounds?

2. Try listening with the funnel in different positions. Where is the best place to put the stethoscope for the sounds to be loudest? What does this tell you about the position of the heart in the chest?

3. With a water-based felt pen draw a cross where you think the heart is, or mark the place with a pin through your shirt. Compare the position with those of other pupils in your class.

4. Count the number of thuds in one minute. This will tell you the rate at which your heart is beating.

Questions for class discussion

1. There is quite a lot of variation between people in the rate at which the heart beats, but for a person at rest it is generally about 70 beats per minute.

Assuming this figure, how many times does the heart beat
a. in one year
b. in 60 years?

2. If you listen extremely carefully, you will notice that each sound from the heart really consists of two thuds one after the other, the first being louder than the second. The thuds are caused by the closing of the valves inside the heart. Why do you think there are *two* thuds?

The arteries

Every time the heart beats, approximately one cupful of blood is forced out into the arteries. This starts a wave of pressure which travels quickly along the arteries.

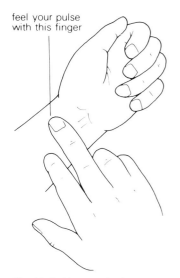

feel your pulse
with this finger

Certain arteries are situated close to the skin, and if you put your finger on one of them you can feel the pressure wave as it passes by. We call this *feeling your pulse*.

Experiment 16.5   Feeling your pulse

1. Gently place the middle finger of your right hand on the wrist of your left hand in the position shown in Fig 16.8. Can you feel a regular throb?

Move your finger around until you find the right place.

2. Count the number of throbs in one minute. This should be the same as the number of heart sounds per minute.

Questions for class discussion

1. If a person loses a lot of blood, for example in an accident, his pulse may become very weak. Why do you think this is?

Fig 16.8 How to feel your pulse

2. If a bandage is tied tightly round your arm, you may not be able to feel your pulse at all. Why do you think this is?

The capillaries

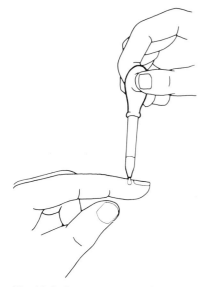

Once an artery has entered an organ it splits up into a large number of narrow capillaries. As blood flows through the capillaries, it brings to the cells all the substances they need, including oxygen from the lungs and dissolved food from the gut. At the same time it takes away any substances which they do not want, such as carbon dioxide.

How does oxygen pass from the blood to the cells? The answer is by *diffusion*. Diffusion is explained in *Physics 11–13* (p. 135). Carbon dioxide passes from the cells to the blood in the same way.

Experiment 16.6   Looking at blood flowing through capillaries

1. With a pipette put a drop of clove oil or cedar wood oil on to the skin at the base of one of your finger nails as shown in Fig 16.9. This will make the skin more transparent.

Fig 16.9 Putting a drop of clove oil or cedar wood oil on your skin will make the capillaries show up under the microscope

2. Put your finger under a microscope (low power) with a strong light shining down on it from above. Can you see any capillaries? Squeeze your finger so as to force blood into it. Does this make the capillaries easier to see?

3. Your teacher may show you the webbed foot of an anaesthetised frog, or the tail of a young fish or tadpole, under the microscope. If so, notice the branching pattern of the capillaries (Fig 16.10). Can you see disc-shaped objects passing along the capillaries? They are *red blood cells* (Fig 16.11). They contain a red substance called *haemoglobin* (pronounced he-mo-globin) which gives blood its characteristic colour. Haemoglobin is the part of the blood which carries the oxygen.

Fig 16.10 Photograph of capillaries of a mammal taken down a microscope

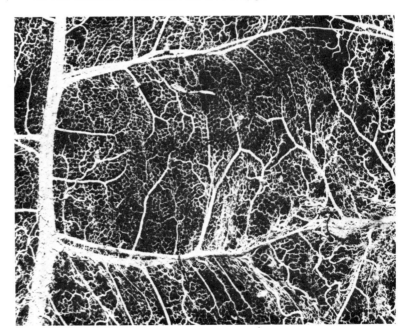

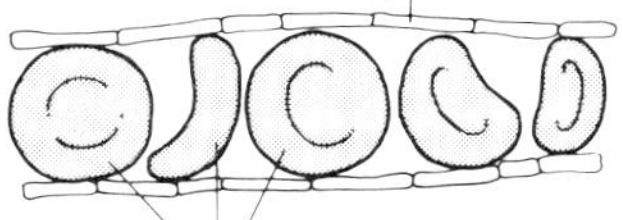

capillary wall

red blood cells

Fig 16.11 Red blood cells in a capillary

Questions for class discussion

1. Capillaries have extremely thin walls – much thinner than either the arteries or veins. Why do you think this is important?

2. If you were to lay out a person's capillaries end to end, they would extend right round the world two and a half times! Why do you think we need so many?

3. An important function of blood, apart from carrying oxygen and other substances, is to spread heat evenly round the body. What process produces the heat, and which organs produce a particularly large amount of it? Why is it important that one part of the body should not be much hotter or colder than another?

The veins

After the blood has been through the capillaries, it is collected up into the veins which take it back to the heart.

Experiment 16.7 Looking at veins in your arm
Work in pairs. Get your partner to hang his arm downwards for a minute or so. Then wrap a bandage round the upper part of his arm just below the shoulder: do not make it so tight that it hurts! The bandage will make the veins near the surface in the lower part of his arm stand out clearly, and this is further helped if he clenches his fist (Fig 16.12). The arteries are deeper down so you will not be able to see them.

When the veins are visible, lightly ink in their course with a water-based felt pen.

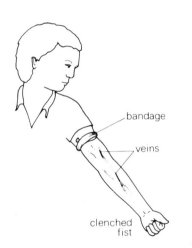

Fig 16.12 Making the veins show up in your arm

Questions for class discussion
1. Why do you think tying the bandage round the arm helps to make the veins show up?

2. Some people have clearer veins in their arms than others. Why do you think this is?

Homework assignments
1. Draw an outline of the human body viewed from the front.
a. by means of an arrow and the letter H, indicate the position of the heart.

b. by means of an arrow and the letter X, indicate a place, a long way away from the heart, where you can feel that your heart is beating.

2. What do we call blood vessels:
a. which carry blood away from the heart;
b. which carry blood towards the heart;
c. which supply oxygen through their thin walls to our cells.

3. Assuming that a person's heart beats at a constant rate of 70 times per minute, how many beats does it undergo in the course of one year?

4. Approximately 150 cm³ of blood are expelled from the heart every time it beats. How many litres are expelled from the heart in
a. one minute
b. one day
c. one year?

5. The human body contains approximately 5 litres of blood altogether and about 150 cm³ of blood are expelled from the heart at each beat.
a. What percentage of the total volume of blood in the circulation is pumped out of the heart every time it beats?
b. Assuming that the heart beats 70 times per minute, how long does it take for all the blood in the circulation to pass through the heart?

Changes during exercise

We know from everyday life that if we exert ourselves we breathe more quickly. In this way, we take more air into our lungs and thus supply our muscles with more oxygen for getting energy. Does the same kind of speeding up occur in our circulation?

Experiment 16.8   The heartbeat during exercise
1. Sit down for 2–3 minutes, and then count the number of your heart beats per minute (see p. 229).

2. Now take some strenuous exercise. Your teacher will tell you what kind of exercise to take and how long to take it for. Then sit down again and count the number of your heart beats per minute.

By how much has your heart beat rate increased? There are several ways of expressing the answer to this question. Choose the way you consider to be the best.

Questions for class discussion
1. This experiment has shown that the heart beats much faster immediately after exercise. Why is it important that this should happen?

2. When the exercise begins, how do you think the heart knows that it should beat faster? Think of as many theories as you can: you will not be asked to prove them!

Keeping cool

When you take exercise your muscles work faster and make more heat. As a result you get hotter. The body responds to this by cooling itself. One way of achieving this is by *sweating*.

Experiment 16.9   To show sweating
You will need a piece of dry cobalt chloride or thiocyanate paper approximately 1 cm × 2 cm. The paper is blue when dry, but turns pink when moistened with water.

1. With Sellotape stick the paper to the skin on the back of your hand. Note the time, and observe the paper at intervals. Does any part of it turn pink? How long does it take for the first trace of pink to appear? Does the entire piece of paper turn pink eventually, and if so how long does it take? What substance must be present in sweat?

2. Repeat this experiment immediately after taking strenuous exercise. How long does it take for the paper to turn pink this time? Explain the difference.

Questions for class discussion

1. After taking strenuous exercise you often feel thirsty. Why do you think this is?

2. Why do you think sweating helps to cool the body? To help you answer this question try this simple experiment. Lick the back of one of your hands, and leave the other one dry. Gently wave both hands in the air. Which one feels cooler? What happens to the moisture on the back of the hand which you licked?

3. If there is a spell of very hot weather, a head teacher may tell the pupils to eat plenty of salt. Why do you think this is done? (Hint: what does sweat taste of?)

Keeping warm

Sometimes we need to keep our bodies warm. When do we need to do this? Humans keep themselves warm mainly by wearing clothes. However, other mammals are kept warm by their *hair*. The hair reduces heat loss, in other words it helps to *insulate* the body. *Fat* under the skin also helps to insulate the body. In birds insulation is achieved by the *feathers*.

As a result of this insulation, much of the heat which is produced by respiration can be kept inside the body. For this reason, mammals and birds are described as *warm-blooded*. Warm-blooded animals can keep their body temperatures constant, whatever the surrounding temperature: if it is cold they generate extra heat and keep it inside the body; if it is hot they let some of the heat out.

Other animals, such as fishes and reptiles, are *cold-blooded*. They have no insulation devices, and their body temperatures are the same as the surrounding temperature.

The only way in which a cold-blooded animal can control its body temperature is by making sure that it is in a place where the surrounding temperature is agreeable.

Questions for class discussion

1. Lizards are described as cold-blooded, but if you pick up a lizard it may feel warm. Explain.

2. What are the disadvantages of being cold-blooded, and what are the advantages of being warm-blooded?

3. In the Arctic and Antarctic there are many species of mammals and birds, but very few amphibians and reptiles. Suggest reasons for this.

Homework assignments

1. In the following statement, which words are *correct*:
During exercise a person produces more
carbon dioxide/oxygen/carbohydrate/heat/energy/
sugar/sweat.

2. A schoolboy dashes up three flights of stairs to collect a book he has forgotten. At the top of the stairs he notices that his heart is beating very fast.
a. Put forward an explanation to account for the change in the rate of his heart beat.
b. List other changes in the functioning of his body which he might detect after this bout of exercise.

3. A girl rushes to catch the bus. By the time she gets to her seat she is feeling very hot. Where and how did the heat arise in her body?

Later in the day the girl competed in a two-mile cycle race, at the end of which she felt in need of a long drink. Explain why she felt thirsty.

4. A scientist measured the amount of oxygen taken in by a man at rest and when active. She then repeated the experiment with a locust. She expressed her results in cubic centimetres of oxygen consumed per gram of body mass per hour. Her results are shown in Table 16.1.

Table 16.1

	Oxygen consumption ($cm^3/g/h$)
Man at rest	60
Man running	905
Locust at rest	0.63
Locust flying	15

a. Why do you think the figures for the man are higher than those for the locust?

b. Why does oxygen consumption increase during activity?

c. Suggest two changes which took place in the man's body while he was running which enabled him to take in more oxygen.

Background reading

In this passage a former Olympic Gold Medallist discusses the importance of training in improving the circulation.

The athlete strengthens his heart by training – it is mostly muscle and responds in the same way, thereby becoming more efficient. This can be proved by taking someone's pulse rate before he starts a training programme, and continuing to measure it as training proceeds. Over a period of weeks it will be found that the rate of the heart beat, when the person is completely at rest, gradually decreases. This resting pulse (usually measured immediately after waking in the morning) will be seventy or eighty beats per minute for the average man or woman, but with training it may well drop to sixty or below. My own resting pulse, when really fit, has been as low as thirty-eight. This is because the heart, being stronger than before and capable of pumping more at each beat, does not have to work as hard to keep the body supplied with oxygen.

The other effect of training is to improve the circulation to the working muscles with the result that more capillary vessels develop in the muscles that are being used most. These capillaries are not in use when you are at rest, but in an emergency, when more oxygen is needed, they open up to allow more blood to flow. This means that you can 'switch on' to a higher rate of exercise when you really need to. Of course this would not be any good unless your heart had also strengthened so that it could deal with this extra effort.

(From *Naturally Fit* by Bruce Tulloh, published by Arthur Barker Ltd.)

Questions

1. Why is a trained athlete's heart able to pump more blood at each beat?

2. Why is it best to take the resting pulse first thing in the morning?

3. What other organs, besides the heart, improve with training? Explain your answer.

4. What other changes might occur in your blood circulation as you get fitter?

Summary
1. *Respiration* is the process by which energy is released in living things.

2. The raw materials of respiration are food and oxygen, and the products are carbon dioxide and water.

3. We obtain oxygen and get rid of carbon dioxide by *breathing*. In this process air passes in and out of the *lungs*.

4. Oxygen is carried from the lungs to all parts of the body by the *circulation* which consists of the *heart* and *blood vessels*.

5. When we exert ourselves we breath faster and more deeply and the heart beat rate increases.

6. During exercise heat energy is released, but the body keeps cool by *sweating*.

7. If it is cold, heat produced by respiration is kept inside the body.

Chapter 17 **Responding to stimuli**

Suppose your friend puts a drawing pin on your chair and you sit on it. You would jump up quickly! (Fig 17.1). The pricking is called the *stimulus*, and your *response* is to jump up.

Responding to stimuli is important to all living things. It helps to protect them from harm and in some cases it may guide them towards food.

Experiment 17.1   Observing the response of blowfly maggots to light

This experiment should be performed in a darkened room.

1. Lay a sheet of rough brown paper, approximately 30 centimetres square, on a table, and place a lamp on one side as shown in Fig 17.2. The lamp should be switched off.

2. Put one maggot in the centre of the brown paper, and observe its behaviour.

3. Switch on the lamp and watch the maggot. Write down what it does.

4. Repeat the experiment six times. Describe the maggot's response each time.

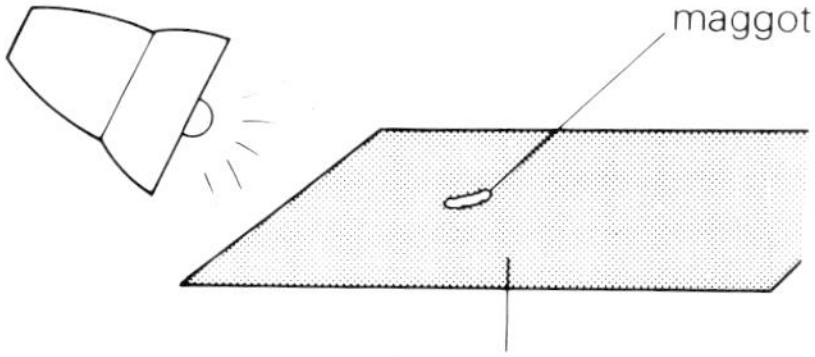

Questions for class discussion

1. Why was it necessary to repeat the experiment six times? Do you think it should have been repeated more times than that?

2. It is possible that the stimulus to which the maggot responded was *heat* from the lamp, rather than light. How could you find out if this is true?

Fig 17.1 Ouch!

maggot

brown paper

Fig 17.2 How to investigate the response of a blowfly maggot to light

3. The eggs from which blowfly maggots develop are normally laid in dung or rotting material. Why is the behaviour which you observed in Experiment 17.1 useful to them?

4. Suggest other stimuli to which maggots might respond. In each case, devise an experiment to find out if they really do respond to the stimulus.

Homework assignments

1. The pictures in Fig 17.3 show five situations, each involving a stimulus and a response. Describe the stimulus and the response in each picture.

2. Given a corked test-tube, some black paper, a lamp and a tinful of maggots, describe fully how you would set about proving that maggots prefer darkness to light.

3. It has been said that an African clawed toad finds its food by smell. Describe how you would set up and carry out an experiment to test the truth of this statement.

Fig 17.3 In each of these pictures, what is the stimulus and what is the response?

(a)

(b)

(c)

(d)

(e)

4. A scientist carries out an experiment on the behaviour of woodlice. He places twelve woodlice in a dry dish at one end of which there is a piece of damp bark from a tree. Every five minutes he counts the number of woodlice that are in the open part of the dish and, by subtraction, he works out how many are underneath the piece of bark. His results are shown in Table 17.1.

Table 17.1

Time (minutes)	Number of woodlice in the open	Number of woodlice under the bark
0	12	0
5	10	2
10	7	5
15	5	7
20	3	9
25	4	8
30	0	12

a. Plot these results as a bar chart (histogram).
b. How do you think the woodlice might find their way to the piece of bark?

Human responses

Think of the various things you do during the day. Which ones involve responding to a stimulus? Make a list of the various stimuli you receive, and in each case describe your response. Two examples are given in Table 17.2 to get you started.

Table 17.2

Stimulus	Response
1. I receive a prick on my bottom from a drawing pin	I jump up and let out a cry
2. I am at the starting post and hear the pistol shot	I start running

Look carefully at your list. You will probably find that in some of your examples the response is quick and does not go on for long. This is the case, for instance, when you jump up after sitting on a pin. We will now look at two other quick responses.

Experiment 17.2   Some human responses
Work in pairs, one person acting as the subject.

Blinking
1. The subject should open his or her eyes and look straight ahead.

2. The experimenter should suddenly wave his or her hand in front of the subject's eyes. Describe what happens. Why is this response useful?

3. The experimenter should repeat the stimulus approximately once every three seconds. Does the subject blink each time?

The pupil response
1. The subject should sit in a dimly lit room with his or her eyes open.

2. The experimenter should observe the two main parts of the subject's eye, namely the *pupil* and the *iris* (Fig 17.4).

3. The experimenter should now hold a torch in front of the subject's eye, and suddenly switch it on.
   What happens to the pupil?
   Why is this response useful?

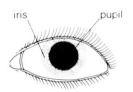

Fig 17.4 The human eye

How are responses brought about?
For a response to occur three things are needed.

1. *Receptors*: these are structures which receive the stimulus.

2. *Nerves*: these are like cables and they conduct messages from one part of the body to another.

3. *Muscles*: these are attached to the *skeleton* and when they receive a message from a nerve they contract and make the body move.

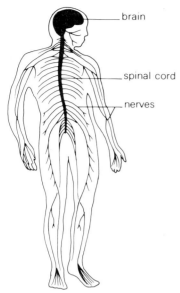

Fig 17.5 The main parts of the human nervous system

Clearly our nerves are very important in bringing about responses. The human nervous system is shown in Fig 17.5. It includes two main parts: the *brain* and the *spinal cord*. The brain is enclosed within the skull in the head, and the spinal cord runs down the middle of the backbone ('spine'). The brain and spinal cord are connected to our skin and muscles by nerves.

Now suppose you get into the starting position for a hundred metre sprint. At the sound of the pistol your ears are stimulated and a *message* is sent through your nervous system to certain muscles in your legs. The muscles then contract and you start running (Fig 17.6).

What part does the brain play?

Look at Fig 17.6. Notice that the message travels from the ears to the muscles by way of the brain.

Our brain is like an immensely complicated telephone exchange with millions of connections. It makes sure that messages are sent to the right muscles so that the correct response is given. The word we use for this is *co-ordination*.

The brain is also responsible for thought, memory and intelligence.

Questions for class discussion

1. Write down a *stimulus* and describe the *response* in each of the following situations. Try to think of a different stimulus in each case.
a. The end of a school lesson.
b. Lunch time.
c. Reading a book.
d. A fire practice.
e. A game of football.

2. What part does the brain play in bringing about the responses which you have given in your answers to the previous question?

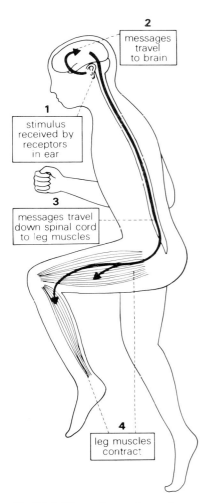

Fig 17.6 The pathway through which messages travel in bringing about a response

The speed of nervous messages

Messages travel very rapidly through the nervous system, and this enables us to respond quickly to stimuli. You can get an idea of the speed by measuring how long it takes for you to respond to a stimulus. This is called your *reaction time*.

Experiment 17.3 Comparing people's reaction times

Work in pairs, one person acting as the experimenter, the other as the subject.

1. Experimenter: hold a ruler as shown in Fig 17.7 so that zero is level with the edge of a table top.

2. Subject: place your hand on the table so that you are ready to catch the ruler.

3. The experimenter should now let go of the ruler and the subject should catch it as soon as possible after it begins to fall.

4. Note the position along the ruler at which the subject catches it. Read off the distance to the nearest millimetre.

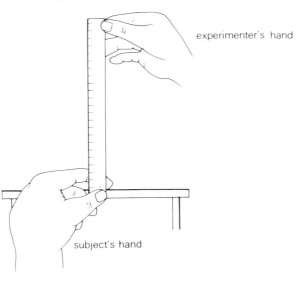

experimenter's hand

subject's hand

Fig 17.7 How to compare people's reaction times

5. Repeat this test with the same subject twenty times, and record the distance each time.

6. Write down the shortest distance achieved by each member of the class.

Questions for class discussion

1. What was the shortest distance, and the longest distance, which was recorded in the class? (This is a good example of human variation: see p. 14–18.)

2. Did your reaction time get shorter with practice? If it did, how would you explain the improvement?

3. Trace the pathway through which messages travel in bringing about this response. Where are the receptors and muscles?

Receptors

In order to respond to stimuli, an animal has to have receptors. A receptor is a structure which can receive a stimulus and send a message in the nervous system.

The main receptors in the human body are shown in Fig 17.8. Most of them are complex *sense organs*, the eye and ear being two important examples. Each receptor is adapted to receive a certain kind of stimulus. Thus the eye receives light rays, and the ear receives sound waves.

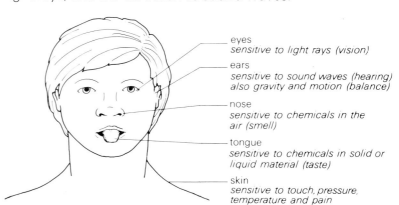

eyes
sensitive to light rays (vision)

ears
*sensitive to sound waves (hearing)
also gravity and motion (balance)*

nose
*sensitive to chemicals in the
air (smell)*

tongue
*sensitive to chemicals in solid or
liquid material (taste)*

skin
*sensitive to touch, pressure,
temperature and pain*

Fig 17.8 The main receptors of the human

244

The eye contains a *lens* which focuses the light rays on a layer of sensitive cells at the back of the eye. The ear contains a membrane, the *ear drum*, which vibrates when sound waves hit it. The ear also contains special receptors sensitive to gravity and motion which help us to keep our balance.

Questions for class discussion

1. An animal's receptors are situated mainly in its head. Why is this an advantage?

2. What kind of receptors in the human body are *not* situated in the head. (Hint: think of the different kinds of stimuli to which your skin is sensitive.)

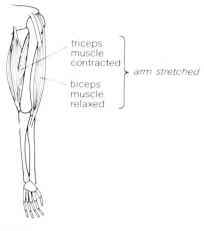

triceps muscle contracted
biceps muscle relaxed
} *arm stretched*

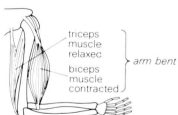

triceps muscle relaxed
biceps muscle contracted
} *arm bent*

Fig 17.9 The muscles and bones involved in bending and stretching your arm

Muscles and the skeleton

An animal must be able to move if it is to respond to a stimulus. This is achieved by muscles. In the human, the muscles are attached to the skeleton. The skeleton is made up of *bones* which are connected to each other at *joints*. When our muscles contract, the bones move smoothly against each other at the joints.

As an example, consider the bending of your arms (Fig 17.9). The elbow joint is like a hinge, and the action of the muscles is rather like opening and closing a door. You can find out about the forces involved in movement from your physics book (*Physics 11–13*, Chapter 9).

Experiment 17.4   Feeling muscles in action

If you put your hand on the surface of the body where a muscle is contracting, it feels hard and a bulge develops.

1. Roll up your sleeve, and stretch out your right arm in front of you.

2. Place your left hand on the upper side of your right arm (Fig 17.10(a)).

245

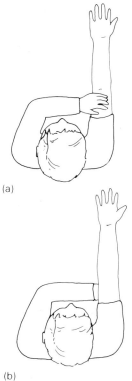

(a)

(b)

Fig 17.10 Feeling your arm
muscles in action

3. Slowly bend your arm. Can you feel a bulge? This is your *biceps muscle*.

4. Straighten your arm, and continue to feel the biceps with your left hand.

5. Get someone to place a heavy object such as a book on your right hand. What does the biceps do? Explain your observation.

6. Keeping your right arm straight, press downwards on a table top.
   What does the biceps do now? Explain your observation.

7. Now place your left hand on the lower side of your arm (Fig 17.10(b)). Can you feel a bulge? This is your *triceps muscle*.

8. Repeat steps 5 and 6 with your hand on the triceps. What does the triceps do at each step? Explain your observations.

Questions for class discussion
1. What happens to the length of a muscle when it contracts? Why does it bulge and go hard?

2. What can you say about the functions of the biceps and triceps muscles?

3. What other muscles in the human body work on the same principle as the biceps and triceps muscles?

Homework assignments
1. What structure, or structures, in the human body:
a. transmit messages from one place to another;
b. are stimulated by light rays;
c. store information in the form of memory;
d. can get shorter?

2. A boy gets into the starting position for a hundred metre sprint. At the sound of the pistol he begins to run.
a. What is the stimulus which makes him start running?
b. Name the receptor which receives the stimulus.

c. Describe the route through which messages travel in the boy's body to bring about the response.

3. Choose a team ball game which you know well. Soccer, netball or hockey would do. Write down the name of the game you choose. Now suppose a member of your team passes the ball to you.
a. Write down the stimulus or stimuli (there may be more than one) which you receive from your team mate before you start responding.
b. What receptor or receptors receive the stimuli?
c. Describe your response.
d. What muscles are involved in the response? (Do not name them but try to explain where they are and how they help to bring about the response.)

Plant responses

Do the plants in your garden respond to stimuli? At first sight you might say no: after all, if you prod a geranium it does not move away. However, plants *do* respond to stimuli. They do so by growing in a particular direction. One of the most important stimuli to which plants respond is light.

Experiment 17.5   Observing the response of a seedling to light

1. Lay a sheet of blotting paper on the bottom of two dishes.

2. Moisten the blotting paper with water.

3. In each dish sprinkle some seeds of mustard or cress, then cover the dish.

4. After the seeds have germinated, put one dish in a window where it will be brightly lit from only one side, and put the other dish in a place where it will be lit equally from all sides.

5. Observe the seedlings at intervals during the next week.

247

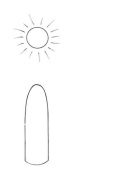

equal growth on all sides of shoot

more growth on the left

Fig 17.11 How a shoot bends towards light

Fig 17.12 A plant bending towards the light

6. After a week make a sketch of one seedling from each dish to show the difference. In each sketch, show the position of the light source.

Questions for class discussion
1. What conclusion can you draw about the way shoots respond to light?

2. Why is this response useful to the plant?

3. What other stimuli might a shoot respond to? In each case devise an experiment to find out if your suggestion is correct.

4. What kind of stimuli do you think a *root* might respond to? In each case devise an experiment to find out if your suggestion is correct.

How does a shoot bend towards light?

The answer to this question is that the shoot grows more quickly on the darker side than on the lighter side. The result is that it bends towards the light (Fig 17.11).

Can you think of any experiments which could be done to show that this is true?

Homework assignment
1. A lady puts a potted plant on a windowsill. After ten days she notices that it has bent over towards the light as shown in Fig 17.12.
a. Suggest two stimuli which might have caused the plant to take on this appearance.
b. What could the lady do to straighten the plant?

Background reading

The champions of hearing, by any standard, are the bats. Bat sounds long went undetected by man because they are pitched two or three octaves above what we can hear. But to a number of bats flying around on a calm, still summer evening – and to the unfortunate moths that can hear them and must try to avoid them – the evening is anything but calm. It is a madhouse of constant shrieking. Each bat sends out a series of screams in short pulses, each lasting a hundredth of a second.

What is important to the bat is not its sound but its echo. Bouncing off obstacles like trees, walls and flying insects, the echoes of its cries keep the bat informed of things in its way and food on the wing. This echo-location device, which acts much like the sonar employed in submarines, has evolved in different ways. Some bats send out a wide, scattered beam, others a narrow one that can be changed in its direction and thus used as a scanning device. We know that echo-location involves the bats' ears, mouths and, in some species, noses, because if any of these are blocked, the bats fly 'blind'. But how the bats' ears and brains process the information they receive from the echoes is still a mystery – their hearing equipment must be complex.

Whales use a similar 'sonar' system in the water. It has long been known that whales could hear, and that they could make sounds too. British sailors called one particularly talkative species, the arctic beluga whale, the 'sea canary'. But the full story was not revealed until World War II, when hydrophones, developed for the detection of submarines, picked up the amazing variety of underwater noises produced by whales and dolphins. We know that at least some whales can transmit ultrasonic sounds as high-pitched as those of bats. But we still do not understand how they produce these sounds, since they have no vocal cords. We also know that whales use echo-location as bats do for avoiding obstacles and for finding prey, and that they have their own vocal 'language'.

(From *Animal Behaviour* by Niko Tinbergen, published by Time-Life International.)

Questions
1. In the process by which a bat avoids an obstacle, what is the stimulus and what is the response?

2. Why did bat sounds go undetected by humans for a long time?

3. In the vocal language of whales and dolphins what kinds of things might the animals be saying to each other?

Summary

1. Human beings and other living things respond to *stimuli*.

2. A response is brought about by a *message* which passes rapidly from a *receptor* to a *muscle* by way of *nerves*.

3. The *brain* ensures that the right responses are given and it also enables us to think and remember things.

4. Movement is brought about by the contraction of *muscles* which are attached to the *skeleton*.

5. Plants respond by growing towards, or away from, a stimulus. For example, shoots grow towards light.

Chapter 18 # Human reproduction and development

Human beings can produce offspring only by means of sexual reproduction. This requires two partners, a *male* and a *female*. Both produce sex cells or *gametes* (see Chapter 6). The female's gametes are *eggs*, and the male's are *sperms*. Eggs are produced by the *ovaries* and sperms by the *testes*.

Puberty

The ovaries and testes do not start releasing eggs and sperms until the person is about twelve or thirteen years old. The time when this happens is called *puberty*. Puberty is brought on by chemical substances called *sex hormones* which are produced by the ovaries and testes and pass round the body in the bloodstream.

Various other changes take place at puberty. In boys the penis grows larger, the voice breaks, and hair starts growing on the body. In girls the breasts develop, and approximately every 28 days, a small amount of bleeding occurs from the womb. This is described as a 'period'; its proper scientific name is *menstruation* which comes from the Latin word 'menstrualis' meaning 'monthly'.

Menstruation is the first of a whole series of changes which take place in the female's body during the course of about a month. These changes make up the *menstrual cycle* which we shall come back to in a moment.

Women generally stop producing eggs around the age of 45–50. This is known as the *menopause* or the 'change of life'. Men can go on producing sperms until well into their seventies.

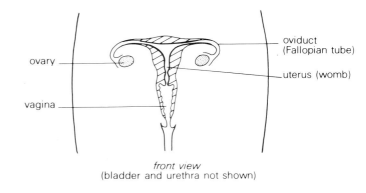

front view
(bladder and urethra not shown)

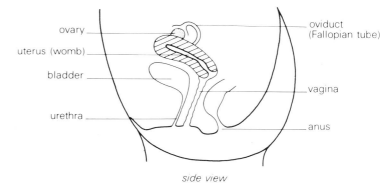

Fig 18.1 The reproductive organs
of the human female. Notice that
the vagina and bladder have
separate openings to the outside.
The urethra carries urine.

side view

The female's reproductive organs

The reproductive organs of the human female are shown in
Fig 18.1.

The eggs are produced by the *ovaries*. There are two
ovaries, one on each side of the body. A tube, called the
oviduct (or Fallopian tube), leads from each ovary. The
oviducts open into a chamber called the *uterus* (or 'womb').
The uterus has a soft inner lining surrounded by a thick
muscular wall. It is here that the baby develops.

From the uterus, a tube called the *vagina* leads to the
outside. Close to the vagina is another much narrower tube
called the *urethra* which comes from the bladder.

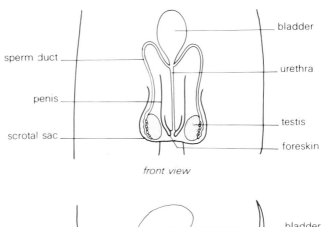

front view

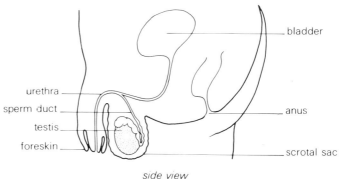

side view

Fig 18.2 The reproductive organs of the human male. Notice that the urethra is connected to the bladder as well as to the sperm ducts; it carries urine as well as sperms.

The male's reproductive organs

The reproductive organs of the human male are shown in Fig 18.2.

The sperms are produced by the *testes*. There are two testes and they lie side by side in a bag called the *scrotal sac*. The testes are thus located outside the main body cavity. Here the temperature is slightly lower than the general body temperature; sperms develop best at this lower temperature.

From each testis, a tube called the *sperm duct* leads towards the lower part of the abdomen. The two sperm ducts join the urethra which runs down the centre of the *penis*. The head of the penis is very sensitive and is protected by the sheath-like *foreskin*.

The menstrual cycle

This is what happens in the menstrual cycle. First menstruation occurs: it goes on for about five days and during that time the lining of the uterus is shed and passes out of the vaginal opening. Then the lining of the uterus heals and thickens and lots of blood vessels develop in it. At the same time, an egg develops in one of the ovaries. About fourteen days after menstruation began, the egg is released from the ovary. The egg enters the upper end of the oviduct. If it is not fertilised within a day or two, it dies. The lining of the uterus continues to thicken and it gets richer in blood vessels. Fourteen days later it is shed and menstruation occurs again.

So menstruation and the release of the egg alternate, with an interval of about fourteen days between each. Menstruation is the first step in preparing the uterus to receive an embryo if the next egg becomes fertilised.

Intercourse and fertilisation

In order to fertilise an egg, the sperms must be placed inside the female's body. This is achieved by *intercourse* (Fig 18.3).

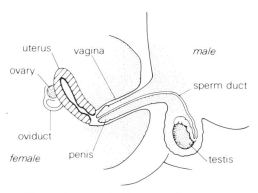

Fig 18.3 In the act of intercourse, sperms are introduced by the male into the vagina of the female

First the penis has to become hard. This is called an *erection*. The male puts his erect penis into the vagina of the female and moves it rhythmically in and out. This stimulates the sensitive penis and after a short while about a teaspoonful of whitish fluid shoots out of the opening of the penis. This process is called *ejaculation*. The fluid is called *semen* and it contains millions of sperms.

The sperms swim through the uterus and into the oviducts. If an egg is present in an oviduct, one of the sperms may fertilise it: the head of the sperm penetrates the egg and its nucleus combines with the egg nucleus (see Chapter 6).

The fertilised egg now divides into a ball of cells which moves down the oviduct to the uterus. Once in the uterus, it sinks into its soft inner lining. The ball of cells is an *embryo*, and the female is now *pregnant*. She will have no more periods, and her ovaries will produce no more eggs, until after her baby has been born.

Development of the embryo

The embryo grows and develops, and it becomes surrounded by a bag of watery fluid. The bag is called the *amniotic sac*, and the fluid inside it is called *amniotic fluid*. It helps to protect the developing embryo from damage.

Fig 18.4 A human foetus in the uterus about three months after the egg was fertilised

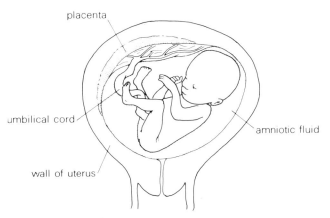

placenta

umbilical cord

amniotic fluid

wall of uterus

By the end of the third month the embryo is approximately 10 cm long, and it looks like a miniature human being. We call it a *foetus*.

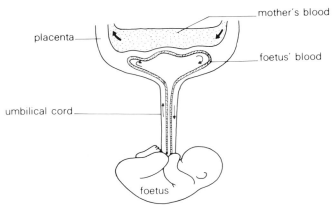

mother's blood

placenta

foetus' blood

umbilical cord

foetus

Fig 18.5 Simplified diagram of the placenta

The foetus is nourished by a structure called the *placenta*. This is shaped like a plate and is attached to the lining of the uterus. The placenta is connected to the foetus by the *umbilical cord* which contains blood vessels from the foetus.

Inside the placenta, blood from the foetus flows very close to the mother's blood (Fig 18.5). As the foetus' blood flows past the mother's blood, it picks up oxygen and food substances and it gets rid of carbon dioxide and nitrogen waste. The foetus' and mother's blood flow very close to each other in the placenta, *but they do not mix*. People differ in the kind of blood they possess: we say they belong to different 'blood groups'. If a mother and her baby belong to different blood groups, their blood might be harmed if allowed to mix together, and that would be fatal. Another reason why the two bloodstreams must not mix is that the mother's blood pressure is much higher than that of the foetus.

The time between fertilisation and birth is known as the *gestation period*. In the human this is approximately nine months. During this time the foetus grows considerably and the uterus expands to make room for it.

It is important that the mother should eat sufficient amounts of the right kinds of food while she is pregnant, and that she should not take into her body anything that might harm the baby. Smoking and alcohol can be harmful to the developing baby. So can germs which get across the placenta. The germs that cause German measles are particularly dangerous: for this reason teenage girls are given an injection to protect them against this disease.

Birth

First the amniotic sac bursts and the amniotic fluid passes out through the vagina. Then the muscles in the wall of the uterus contract powerfully and the baby is forced through the vagina, usually head first.

After the baby has emerged, the placenta comes away from the lining of the uterus and is discharged through the vagina: this is called the *afterbirth*.

As soon as the baby has been born it starts to breathe. The umbilical cord, of no further use, is cut and the scar becomes the baby's *navel* or 'belly-button'.

Fig 18.6 A baby is born

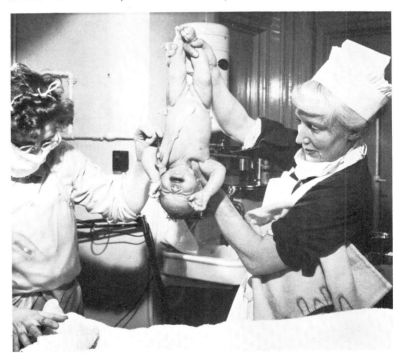

Growth and development

During the first few months of its life the baby can get all the food it needs by sucking the nipples on the mother's

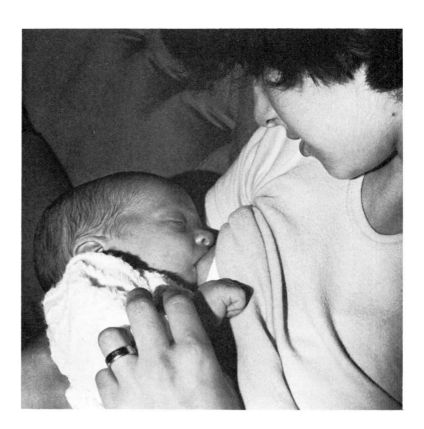

Fig 18.7 A mother breast-feeding her baby

breasts (Fig 18.7). The breasts contain *mammary glands* which produce milk. This is a complete food and contains all the substances necessary for the baby's healthy growth. Alternatively, the baby may be fed on milk from a bottle with a teat.

The child grows rapidly for the first few years, then more slowly. Growth speeds up again during *adolescence* – approximately between the ages of twelve and eighteen years. It then slows down and stops by the age of about 20 years. The different parts of the body do not all grow at the same rate: for example, during the first four years or so, the head grows much more quickly than the rest of the body.

It is important that the growing child should have an adequate diet and lead a healthy life (see Chapters 15 and 19). It is the parents' responsibility to make sure of this.

Questions for class discussion

1. Chapter 15 deals with the various substances which must be included in our diet. During pregnancy the mother has to eat extra amounts of these substances. Why?

2. Why is mother's milk said to be a 'complete food'?

3. The foetus has a gut and a pair of lungs. However, neither are working. Explain the reason for this. When do they start working?

4. An important feature of human development is the care of the young by the parents. Make a list of all the ways in which human parents care for their offspring.

5. The semen of a human male may contain as many as 500 million sperms, and yet only one is needed to fertilise an egg. If only one is needed why are so many produced?

Homework assignments

1. In the human female:
a. Name the structures which produce eggs.
b. Whereabouts does fertilisation normally occur?
c. Which structure receives the male's penis during intercourse?
d. What structure houses the developing embryo?
e. Name the structure which nourishes the developing embryo.

2. In the human male:
a. Name the structures which produce sperms.
b. Whereabouts are the structures located, and why?
c. What passes down the urethra, and when?

3.
a. Why can humans produce babies only after both partners have reached puberty?
b. Why can a married couple have no more children after the wife reaches the age of 45 to 50?

4.
a. What is the period between fertilisation and birth called, and how long does it last?

b. Describe how the developing embryo is protected.
c. Materials are exchanged across the placenta between the mother's blood and the embryo's blood.
(i) Name two needs of the embryo that are met in this way.
(ii) Name two waste substances that leave the embryo in this way.
d. One function of the placenta is to keep the mother's blood separate from that of the embryo. Why is this necessary?
e. In most births, which part of the baby emerges from the mother's body first?

5. Find out, either from books or by asking people who know, when (approximately) each of the following events occurs during human development. Qualify your answers where necessary.
a. A baby starts eating semi-solid food.
b. The first tooth breaks through.
c. The colour of the eyes changes.
d. The child starts crawling.
e. The child starts walking.
f. The child says 'mum'.
g. The child says 'dad'.
h. The first set of teeth is complete.
i. The second set of teeth is complete.
j. A girl can become pregnant.

6. The graph in Fig 18.8 compares the growth of three different parts of the body from birth to the age of 20 years.

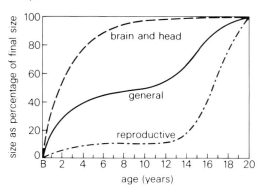

Fig 18.8

a. If the legs were 84 cm long at the age of twenty years, approximately how long were they at five, ten and fifteen years?
b. Why does the head grow so quickly in the early years?
c. Why does the growth of the reproductive organs speed up at the age of twelve years?

Summary

1. The main reproductive organs of the male are the *testes*, *sperm ducts* and *penis*.

2. The main reproductive organs of the female are the *ovaries*, *oviducts*, *uterus* and *vagina*.

3. When *puberty* is reached the testes and ovaries start releasing *sperms* and *eggs*. In the female an egg is usually released once every 28 days, midway between one period and the next. This continues until the age of about 45–50 (the menopause).

4. During *intercourse* sperms from the male are put into the vagina of the female. If an egg is present in one of the oviducts it may be fertilised by a sperm.

5. In the uterus, the *embryo* develops into a *foetus* which is nourished by way of the *placenta* and protected by the *amniotic fluid*.

6. Birth takes place approximately nine months after the egg was fertilised.

7. The child grows and develops through adolescence into an adult, reaching full size at the age of about 20 years.

Chapter 19

Humans and their environment

'Where do you live?' and 'What is your environment?' were two of the questions asked in the first chapter of this book. In this last chapter we return to these questions and consider the ways in which humans are affected by their environment and attempt to control it. Figure 19.1 summarises four important aspects of our environment. We will look at some of these aspects in this chapter.

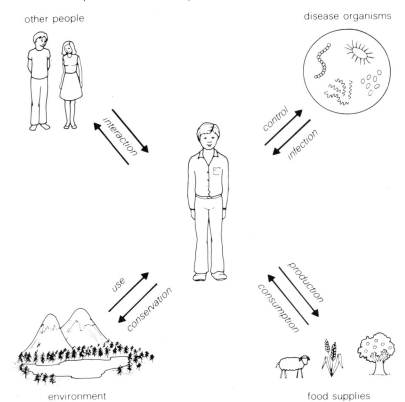

Fig 19.1 Humans and the environment

Disease

Many human diseases are caused by three groups of organisms: bacteria, viruses and fungi. Bacteria and viruses are usually referred to together as germs.

Bacteria

Fig 19.2 Bacteria on the tip of a pin seen through an electron microscope. The bacteria in (d) are magnified 31 250 times

Most bacteria are about one-thousandth of a millimetre long, and are just visible under the light microscope. The more powerful electron microscope shows them better (Fig 19.2).

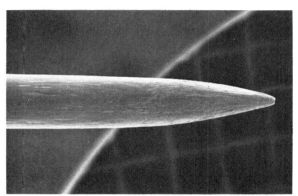

(a) the pin

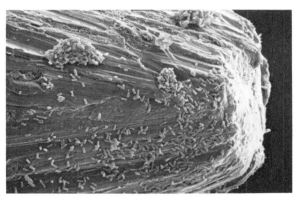

(b) the tip of the pin covered in bacteria

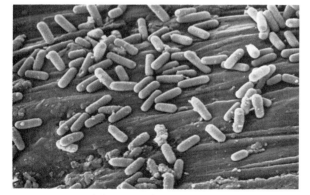

(c) close-up of the bacteria

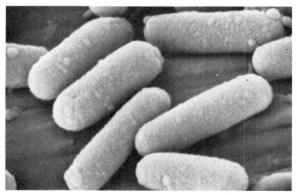

(d) the bacteria magnified greatly

Some bacteria cause human diseases. The Black Death was a bacterial disease which reduced the population of the British Isles by a third in the fourteenth century. It was carried by fleas from rats to people. Many bacteria can reproduce asexually forming highly resistant spores which spread through the air in droplets or by contact. Once inside the body some of the bacteria may be destroyed by the body's natural defence mechanisms. They can now be controlled by various chemical substances including antibiotics. Penicillin is a well known antibiotic which is produced by a fungus.

Many kinds of bacteria play an important part in natural communities. Decay bacteria complete the breakdown of the dead remains of animals and plants, releasing chemicals into the soil for living plants to use. Bacteria are also used by humans. Among other things cheese and yoghurt are made using bacteria.

Viruses

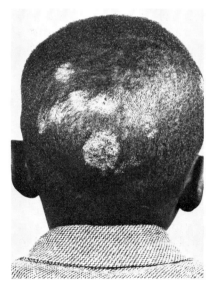

Viruses are much smaller than bacteria. They are too small to be seen under a light microscope. They cannot survive for long outside their host's cells. Drugs are of little use against most virus infections but fortunately we can be protected against some virus diseases by immunisation (see page 268). For example, children are immunised against measles.

Fungi

One of the most common diseases caused by a fungus is ringworm (Fig 19.3). It is an infection of the outer layer of the skin by a microscopical fungus. It frequently forms raw, itching patches between the toes called 'athlete's foot'. It is spread by spores and controlled by various powders and ointments. Infection is less likely to occur if the feet are kept dry, cool and clean.

Fig 19.3 Ringworm infection on the scalp

Health

Good health is achieved not only by avoiding disease; we can also help ourselves to keep healthy in the following ways.

1. Eating the right food

One-third of the world's population does not have enough to eat, and another third has a diet which consists of the wrong kinds of food. Carbohydrates, fats, proteins, vitamins and minerals must be present in adequate amounts, especially in the diet of growing children (see Chapter 15).

In the richer countries the problem is often that people eat too much food. If we eat more than our bodies can use we get fat. Doctors have found that people who overeat are more likely to suffer from diseases of the heart and arteries than other people.

2. Taking exercise

Muscles which are not used regularly become flabby and weak. The muscles of the heart and arteries are no exception – they need to be exercised. Regular exercise reduces fat deposits and increases blood flow.

3. Getting enough sleep

Sleep provides an opportunity for the body to recover from the day's activites. Cutting down on sleep can cause irritability and perhaps increase the risk of disease.

4. Caring for the teeth

Very few adults have perfect teeth and gums. By the age of five, one in every four children in Britain has already had at least one tooth extracted, and two-thirds of all eight-year-olds have had extractions (Fig 19.4). The cause of most dental decay is *plaque*, a mixture of food-remains and bacteria which coats the teeth. The bacteria break down the

Fig 19.4

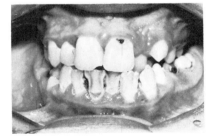

(a) Set of bad teeth

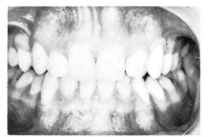

(b) Set of good teeth

food remains releasing acid which attacks the teeth and gums. These bacteria prefer sugars, so eating sticky, sweet foods encourages them. There are four main ways of reducing tooth decay and gum disease:

a. Reduce the number of sticky, sugary foods and sweet drinks that you consume.
b. Brush your teeth regularly, between as well as on the surface of the teeth. Change your toothbrush as soon as the bristles become soft and bent.
c. Use a fluoride toothpaste or take fluoride tablets if your drinking water has no added fluoride.
d. Visit the dentist regularly to have your teeth and gums checked.

5. Keeping clean

The skin provides a tough, protective layer which keeps out infection. However, if it is broken, bacteria may enter the tissues underneath. Washing the skin regularly will help to reduce the risk of infection.

Dirty skin also attracts external parasites such as fleas and lice. If you have a pet, it is important to check its health regularly; fleas and worms are sometimes spread to humans from other animals such as dogs.

The head louse is particularly common in schools today. The female lays up to 300 eggs in her one month of life. These eggs, called 'nits', are cemented to the base of the hairs. They cannot be brushed out but must be treated with a special shampoo and combed out when they are dead.

6. Not smoking

Some of your friends may have begun to smoke; perhaps your parents smoke. Most adults would agree that, if they had known how harmful cigarettes were, they would not have started smoking. For example it has been shown without any doubt that it is a major cause of cancer. Over a third of cancer deaths are the direct result of smoking. It can also have many other harmful effects on your body. Doctors see so many lives affected by it that the majority have given it up. One of the best ways of protecting your health is to decide not to smoke.

Experiment 19.1   How well do you clean your teeth?

Your teeth are normally covered with a layer of plaque. To show the presence of plaque we can use the red dye erythrosin.

1. Rub a little Vaseline on your lips to prevent them being stained by the dye.

2. Suck a disclosing tablet (containing erythrosin).

3. Wash out your mouth with water.

4. Examine your teeth with a mirror and make a drawing in your notebook showing the areas of plaque (stained red).

5. Now brush your teeth thoroughly.

6. Suck a second erythrosin tablet.

7. Make a second drawing to compare the areas of plaque after brushing.

Questions for class discussion

1. Which parts of your teeth were stained red? Which method of brushing is most likely to remove plaque from these areas: side to side, up and down or a circular movement?

2. The amount of stain will vary considerably from person to person. How many explanations can you suggest for this?

3. If toothpaste is tested with a suitable indicator, it can be shown that it is alkaline. How does this help to reduce decay?

4. Carry out a dental survey in your class. Beside each name note the following information:
a. Number of visits to the dentist in a year.
b. Number of times the teeth are cleaned each day.
c. Brand of toothpaste used. Does it contain fluoride?
d. Number of sweets eaten each day.
e. Number of teeth.

f.   Number of extracted teeth.
g.   Number of teeth with fillings.

What conclusions can you draw from the results?

5. There is strong evidence to suggest that cigarette smoking causes lung cancer. Some of the evidence has been obtained from a study of British doctors. The death rate from lung cancer among doctors who smoke is 1.26 per thousand: that is over 30 times greater than the death rate for non-smokers. The results of giving up smoking are shown in the graph in Fig 19.5.

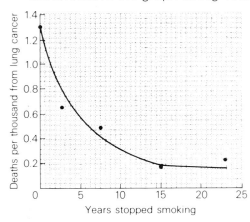

Fig 19.5 The effect on the death rate of doctors of giving up smoking

Describe in your own words what the graph shows about
a.   the results of giving up smoking
b.   the connection between smoking and lung cancer.

Community health

Although many things can be done to avoid disease, we all depend on the community in which we live for our continued health. Here are four things which the community provides.

Immunisation
Many infectious diseases cannot be caught twice. After the

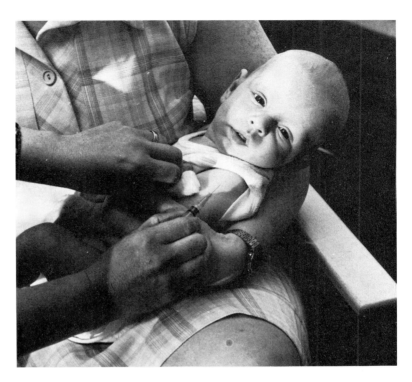

Fig 19.6 A child being immunised

first infection the body is able to defend itself against the germs. In the process of immunisation we are given a small dose of live or dead germs. When these get into the blood, our bodies produce defence chemicals which protect us against the disease in future. In this country, children are immunised against several infectious diseases (Fig 19.6).

Clean food

All fresh food will contain living organisms such as bacteria. If it is kept warm, these bacteria will multiply and make the food decay. They may also produce poisons or *toxins*. To prevent fresh food from going bad it must be preserved by smoking, salting, pickling or drying. Freezing also preserves food but it does not kill the bacteria; it only prevents them from multiplying and feeding. Food which has been re-frigerated should not be re-frozen after being thawed; the

Fig 19.7 Hands should be thoroughly clean if food is to be handled

bacteria may have increased to a dangerous level. When handling food it is particularly important to wash your hands thoroughly (Fig 19.7).

Clean water

The water which we drink may well have been used by several other people beforehand. London water comes from the Thames, which has already supplied Oxford and other towns further up the river. This is made possible by sewage treatment plants (see Fig 19.8). Solid particles separate out in large settling tanks and the liquid is then passed through filter beds rich in decay bacteria before returning to the river. Finally downstream the water is treated with chlorine, a powerful germ killer, before being piped to cities, towns and villages.

Clean air

In the last 30 years the cities and towns of Britain have become much cleaner places to live in because much of the smoke has been removed from the atmosphere. You may have heard of 'smog' – a mixture of fog and smoke – which used to cover London and other large cities for several days at a time. London had a particularly bad smog in 1952 which caused thousands of deaths.

Fig 19.8 Sewage treatment plant Norwich. The large round areas are the filter beds

Fig 19.9 Atmospheric pollution

Homework assignments

1.
a. Why does food decay?
b. List the methods available to preserve food.

2. Prepare a short talk on one of the following:
Preventing tooth decay
The dangers of smoking
Immunisation

3. Design a poster to help prevent the spread of the head louse in schools.

4. Name a human disease which is caused by each of the following organisms, and describe briefly   (i) how it spreads and   (ii) how it can be controlled.
a. A virus   b. A bacterium   c. A fungus.

Pollution

Look at the view through the window of the room in which you are sitting. What do you see? What effect have humans had on this view?

Many different things affect your environment; for example the lie of the land, the weather and the type of soil. But by far the most important factors are humans themselves. Almost every view we see is influenced by our own activities. In particular, when humans release substances into the environment, they may cause *pollution*. An example of atmospheric pollution is shown in Fig 19.9. The effect of oil pollution in the sea is shown in Fig 19.10.

Here is another kind of pollution. The insecticide DDT was first introduced in 1945 and was an immediate success in controlling the head louse in American soldiers in Italy. It was largely responsible for eliminating the malaria-carrying mosquito from Sri Lanka.

The discoverer of DDT was given a Nobel prize, one of the highest scientific awards. For the next twenty years DDT was widely used on crops, animals and even on humans. However DDT can be harmful as well as helpful. Suspicions were first aroused by the gradual disappearance or reduction of certain species of birds. After the spraying of a Californian lake against midges, no grebes (a fish-eating bird) hatched out. The peregrine falcon was almost eliminated as a breeding bird in Britain. Scientists gathered information which suggested that DDT might somehow be responsible, and DDT was banned or restricted in many countries. The bird populations recovered – but malaria returned to Sri Lanka! The explanations for the harmful effects of DDT are clearer now. It seems that three factors are involved:

1. DDT does not break down to simpler chemical substances very easily. In spite of the ban on its use, all organisms now contain traces of DDT or its derivatives.

2. DDT is passed from one organism to another in a food chain, and becomes more concentrated at each level. In an estuary, for example, the concentration in the water is only 0.000 05 parts in every million parts of water (ppm). This 'rises' to 21 ppm in the top carnivores.

Water	$\longrightarrow$	Producers	$\longrightarrow$	Herbivores	$\longrightarrow$	Carnivores	$\longrightarrow$	Top Carnivores
0.000 05 ppm		0.04 ppm		0.3 ppm		3.7 ppm		21.0 ppm

3. The high concentration of DDT sometimes killed the animals but often it affected them in other ways. In the case of the peregrine falcon, DDT interfered with the formation of the egg shell so that the eggs frequently broke when they were incubated.

Fig 19.10 Bird death caused by oil pollution of the coast

high pollution

no lichens – only *Pleurococcus*

crusty lichen

flat leafy lichen

upright leafy lichen

clean air

shrubby lichen

increasing pollution

Fig 19.11 Lichens and pollution

So pollution may have most unexpected consequences for natural communities. We cannot hope to eliminate pollution of our environment but we can reduce it to a minimum and constantly check its effects on living organisms.

Experiment 19.2 How much pollution is there in your environment?

The following simple tests can be used for air and water pollution. If you can compare places near cities or towns with places in the country, you will get some idea of the effects of pollution on your environment.

Air pollution

1. Remove a needle from an evergreen tree such as a pine. Pull it through a damp folded filter paper. Does it leave a mark on the paper?

2. Examine tree trunks, roofs and gravestones. How many different kinds of lichens can you find? Figure 19.11 shows you some of the species you may find, and the levels of pollution which they indicate.

Water pollution

1. Take a clean glass container and fill it with water from a pond or stream near your school. Avoid floating vegetation and stirred-up mud. Leave it to stand for a few minutes, then see how clear it is. Is it coloured? Does it smell? Use a universal indicator paper to test the acidity of the water sample. If you find that the value is below 5, the water is unduly acid and is quite likely to be polluted.

2. Use a pond net to collect animals from beneath the stones or mud. The variety and number of animals present can indicate the amount of pollution (see Fig 19.12).

Questions for class discussion

1. If the rivers or lakes in your area are polluted, what has caused the pollution? What steps could be taken to reduce or eliminate the pollution?

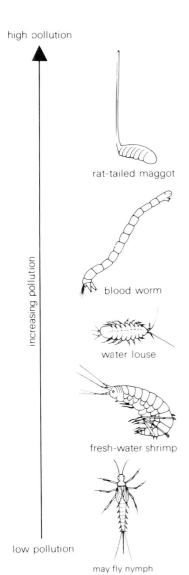

high pollution

increasing pollution

rat-tailed maggot

blood worm

water louse

fresh-water shrimp

low pollution

may fly nymph

Fig 19.12 Some indicators of pollution in fresh water

2. What are the sources of air pollution? How could they be reduced?

3. What harmful effects could air pollution have on animals and plants? List as many effects as you can.

Conservation

The aim of conservation is to maintain a balanced environment. The increasing human pressure on our wildlife has led to many attempts at conservation. These can be roughly divided into efforts to save individual species and attempts to preserve whole communities in their natural habitats. Where possible, the latter is the best policy since it is likely to be a more permanent solution.

The Hawaian Goose or Né-né (Fig 19.13(a)) is a remarkable and successful example of a species saved from extinction. In 1950 the total population in Hawaii was believed to be under 50 and falling rapidly, but 3 birds were brought to England to the Wildfowl Trust Centre at Slimbridge in Gloucestershire. After a successful breeding programme, large numbers of geese were returned to the wild in Hawaii and the bird is no longer in immediate danger.

The Osprey (Fig 19.13(b)) was a common bird in Scotland well into the last century. It was persecuted by land owners, egg and skin collectors until it was eliminated from Britain. The last pair probably nested in about 1916. However, from 1955 to 1958 a pair of ospreys returned to Scotland and attempted to nest at Loch Garten in Invernessshire. The Royal Society for the Protection of Birds established a local nature reserve and three young were reared in 1959. After a number of years of successful breeding at Loch Garten, ospreys are now nesting at about twenty different places in Scotland.

These and many other successes in conservation are the result of the work of large organisations such as the Nature Conservancy Council, Wildfowl Trust, World Wildlife Fund and Royal Society for the Protection of Birds. If you want further information, ask your teacher to give you their addresses so that you can write to them. You can play your part by careful observation and recording of animal and

275

Fig 19.13 (a) The Hawaiian Goose or Né-Né   (b) The Osprey

plant species. Remember to follow the Country Code and avoid damaging the environment. The six main rules of the Code are

1. Do not drop litter
2. Guard against fire
3. Fasten all gates
4. Keep to the paths
5. Avoid damaging fences, hedges and walls
6. Protect all wildlife – take notes, photographs or drawings rather than specimens and never interfere with any bird's nest.

Make sure that the world which your children inherit is a more attractive place to live in than the present one.

Fig 19.14 A notice encouraging people to protect the countryside

Homework assignments

1. Explain what is meant by
a. pollution   b. conservation.

2. What evidence would you look for if asked to say whether or not a pond was free from pollution?

3. Human activities can drastically change natural habitats. Give three examples of such activities and their results.

4. Describe three ways in which fresh food can be preserved and explain why each method prevents decay.

5. Choose an animal which you have studied during your biology course and describe the main danger which threatens it in the wild.

6. Man is at the end of many food chains. What special dangers does this involve for him?

Summary

1. Many human diseases are caused by *bacteria*, *viruses* and *fungi*.

2. We can help to keep ourselves healthy by eating the correct food, taking exercise and sufficient rest, looking after our teeth, keeping ourselves clean and not smoking.

3. The health of the community in which we live depends on the provision of clean water and food, and the prevention of the spread of infectious diseases by *immunisation*.

4. There is a delicate balance between humans and their environment. *Pollution* may destroy the balance; *conservation* can help to restore it.

Appendix 1 # Some useful books

Humans

Craig: *Amazing Facts about Your Body* (Transworld)
Gray: *What is Food?* (Macdonald)
Harris: *Human Measurement* (Heinemann)
Lewis and Rubenstein: *The Human Body* (Hamlyn)
Morgan: *Coordination* (Macdonald)
Pick: *The Mind and Body* (Book Club Associates)
Sanderson: *Respiration* (Macdonald)
Tribe: *Reproduction and Growth* (Macdonald)

Mammals

Bang and Dahlstrom: *Field Guide to Animal Tracks and Signs* (Collins)
Hearne: *Care for your Gerbil* (Collins)
Kelly: *Educational use of Living Organisms: Small Mammals* (Hodder and Stoughton)
van den Brink: *Field Guide to Mammals of Britain and Europe* (Collins)

Birds

Fitter: *Bird Watching* (Collins)
Harrison: *Nests, Eggs and Nestlings of British and European Birds; The Nature Trail Book of Bird Watching* (Usborne)
Reade and Hosking: *Nesting Birds, Eggs and Fledglings* (Blandford)

Fish

Muus and Dahlstrom: *Fresh Water Fishes of Britain and Europe* (Collins)

Insects

Chinery: *Insects of Britain and Northern Europe* (Collins)
Higgins and Riley: *Butterflies of Britain and Europe* (Collins)
Step: *Bees, Wasps, Ants and Allied Insects* (Warne)

Flowers and trees

Cook: *The Nature Trail Book of Woodlands* (Usborne)
Darlington: *World of a Tree* (Faber)
Keble Martin: *Concise British Flora in Colour* (Ebury Press
 and Michael Joseph)
Mitchell: *Trees of Britain and Northern Europe* (Collins)
Simmons: *The Life of Plants* (Macdonald)

Other plants

Hunt: *Discovering Botany* (Longman)
Kelly: *Educational Use of Living Organisms: Plants* (Hodder
 and Stoughton)
Lange and Hora: *Mushrooms and Toadstools* (Collins)
Phillips: *Grasses, Ferns, Mosses and Lichens of Great Britain
 and Northern Ireland* (Ward Lock)

General

Allen and Denslow: *The Clue Books: Freshwater Animals*
 (Oxford University Press)
Arnold: *The Young Ecologist* (Ward Lock)
Barrett and Yonge: *The Sea Shore* (Collins)
Blythe: *Web of Life* (Longman)
Comber and Hogg: *Animals in Schools Vol 2* (Heinemann)
Chinery: *The Family Naturalist* (Book Club Associates)
Chinery: *Natural History of the Garden* (Fontana/Collins)
Darlington: *A Zoo in Your House* (Transworld)
Darlington: *Fun with a Magnifying Glass* (Transworld)
Darlington: *Jam-jar and Saucer Gardens: a Garden in your
 Bedroom* (Transworld)
Gilman: *Urban Ecology* (Macdonald Educational)
Nuffield: *Revised Nuffield Biology Teachers' Guide 1*
 (Longman)
Penguin: *Nature Guides (Birds, Fishes, Fungi, etc.)*

Units used in measurement

Length

1 kilometre	= 1000 metres (m)	= 0.621 miles
1 metre	= 1000 millimetres (mm)	= 3.281 feet
1 millimetre	= 1000 micrometres (μm)	= 0.039 inches
1 micrometre	= 1000 nanometres (nm)	

Mass

1 tonne	= 1000 kilograms (kg)	= 0.984 tons
1 kilogram	= 1000 grams (g)	= 2.205 pounds
1 gram	= 1000 milligrams (mg)	= 0.035 ounces
1 milligram	= 1000 micrograms (μg)	

Area

1 hectare	= 100 ares	= 2.469 acres
1 are	= 100 square metres (m²)	= 10 764 square feet

Volume

1 litre	= 1000 cubic centimetres (cm³)	= 1.760 pints

(a) ×20

(b)

(c)

(d)

(e)

(f)

Appendix 3 # Keys to living organisms

Key to plants in aquarium tanks

(drawings not to the same scale)

Microscopic plants	1
Others	2

1. Long strands or filaments — filamentous algae like *Spirogyra* (*a*)

Single green cells — unicellular green algae

2. Free floating near or at the surface — 3

Submerged and often rooted in mud — 4

3. Flat on both sides, rounded leaves about 3 mm across floating on the surface — common duckweed (*b*)

Just below the surface, branched plant, pointed leaves — ivy-leaved duckweed (*c*)

4. Large broad leaves — common pondweed (*d*)

Small leaves, or fronds (leaf-like parts) — 5

5. Overlapping leaves or fronds — Canadian pondweed

Leaves or fronds not overlapping — 6

6. Whorls with unforked branches — stonewort (*f*)

Other whorls — 7

(g)

(h)

7. Whorls with continuously
forked branches and dense
fronds at the top   hornwort (*g*)
Whorls with little branches on
larger branches   water milfoil (*h*)

Key to animals in aquarium tanks

(drawings not to the same scale)

Soft bodied, no hard covering or
shell   1
Shell or other hard outer covering   5

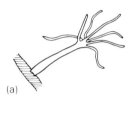
(a)

1. Waving tentacles on the end
of a tube-shaped body, about
10 mm long   *Hydra* (*a*)
No tentacles, worm-like body,
long and flat or round   2

(b)

2. Body not divided into sections
or segments   3
Body segmented   4

3. Body round and thread-like,
free swimming   roundworm (*b*)
Body flat, pointed tail, mouth
underneath, up to 26mm long   flatworm (*c*)

(c)

4. Red in colour, no suckers, tail
often waving out of the end of
a tube sticking up in the mud,
up to 85mm long   *Tubifex*
Suckers at either end   leech (*d*)

(d)

5. Hard external shell of some
sort into which animal can
withdraw. No limbs   6
Animal with jointed limbs   8

(e)

6. Double shell up to 40 mm long   fresh water mussel (*e*)
Single shell   7

(f)

(g)

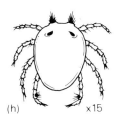

(h)   x 15

(i)

7. Shell a spiral   pond snail (*f*)
    Shell not a spiral   freshwater limpet (*g*)

8. Four pairs of legs 1 to 3mm
    long   water mites (*h*)
    Other number of legs   9

9. Three pairs of legs   10
    More than four pairs of legs   16

10. Wings or wing covers present   adult insects 11
     No wings   insect larvae 12

11. Swims on back, third pair of
     legs fringed with hairs,
     sucking mouth parts   water boatman (*i*)
     Hard wing cover, biting
     mouth parts   water beetle (*j*)

12. Mainly just under the surface
     with breathing tube   mosquito larva (*k*)
     Deeper in the water or on the
     bottom, no breathing tube   13

13. Larva in protective tube   caddis fly
     Not in tube   14

14. Two tail prongs   beetle larva (*l*)
     Three tail prongs   15

15. Long prongs   mayfly nymph (*m*)
     Short flattened prongs   dragonfly nymph (*n*)

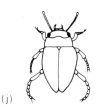

(j)

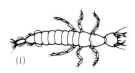

(l)

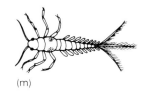

(m)

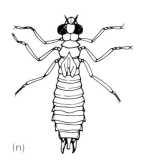

(n)

284

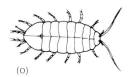

(o)

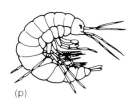

(p)

16. Body more than 5 mm long   17
     Body less than 5 mm long   18

17. Body like a woodlouse, non-
     swimming   water louse (*o*)
     Swims on one side, body
     flattened from side to side   freshwater shrimp (*p*)

18. Pear-shaped body, single
     median eye   Copepods (*q*)
     Squat shaped   water fleas (*r*)

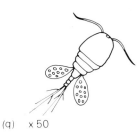

(q)   x 50

(r)   x 50

Key to animals on trees, in litter and in pitfall traps

(drawings not to the same scale)

Animals with no internal skeleton
     or bones   1
Internal skeleton   birds or mammals

1. Body not divided into
     sections called segments   2
     Body divided into segments   4

2. Worm-like body, no shell or
     tentacles, less than 1 mm
     long   nematode worm (*a*)
     Muscular foot, head bearing
     tentacles   3

3. Shell present   snail (*b*)
     No shell   slug

(a)

(b)

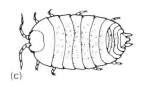
(c)

(d)

(e)

(f)

4.	More than four pairs of legs	5
	Four pairs of legs or less or none	7
5.	More than twenty segments	6
	Less than twenty segments	woodlouse (*c*)
6.	Two pairs of legs per segment	millipede (*d*)
	One pair of legs per segment	centipede (*e*)
7.	No legs, less than 15 segments	insect larvae (*f*)
	No legs, more than 15 segments	26
	Legs present	8
8.	Four pairs of legs	9
	Three pairs of legs	10
9.	Body clearly divided into two parts	spiders (*g*)
	Body not clearly divided	mites (*h*)
10.	Wings present, sometimes hardened	11
	No wings	22
11.	One pair of wings	12
	Two pairs of wings	14
12.	Body about 2.5 cm long, antennae and legs	crane fly (*i*)
	Body about 1 cm long or less	13

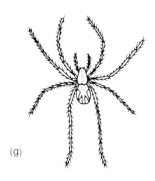
(g)

(h)

(i)

(j)

13. Slender body, long legs,
piercing mouth parts — gnat or mosquito (*j*)
Squat hairy body, short
antennae — other two-winged fly (*k*)

14. Both pairs of wings
membranous — 15
Forewings hardened or
partially hardened — 19

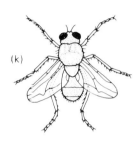

(k)

15. Long thin body, biting mouth
parts, complicated network
of veins on the wings — dragonfly (*l*) or lacewing
Squat body and veins not
complicated network — 16

16. No distinct waist, sucking
mouth parts tucked under
head — greenfly
Distinct waist — 17

17. Large wings covered with
scales, long antennae — butterfly or moth
Small wings — 18

(l)

18. Very narrow waist, long
abdomen with egg-laying
tube — ichneumon fly (*m*)
Hairy body, short abdomen,
often striped — bee or wasp (*n*)

19. Forewing cover partially
hardened, mouth parts for
piercing or sucking — plant bug (*o*)
Forewing cover all hard,
biting mouth parts — 20

(m)

(n)

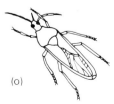

(o)

(p)

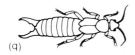

(q)

(r)

(s)

(t)

(u)

20. Forewing covers most of abdomen

ladybird or other beetle (*p*)

Narrow body, dark colour very short wing covers

21

21. Pincers at end of abdomen

earwig (*q*)

No pincers, tail often curled back

coach horse beetle (*r*)

22. Under 6 mm long

23

Over 6 mm long

25

23. Body oval-shaped, with no definite waist, up to 4 mm long

wingless greenfly (*s*)

Body not oval-shaped

24

24. Abdomen 6 segments or less, jumps by spring under abdomen, 1 mm long

springtail (*t*)

More than 6 segments in abdomen, no springs

other groups of small insects

25. Long, rather soft-bodied organisms

insect larvae (*u*)

Definite waist

ant (*v*)

26. Small (2cm long)

whiteworms

Large (up to 15 cm long)

earthworm (*w*)

(v)

(w)

Index